Indigo and Resist Dyeing

Indigo and Resist Dyeing

Performance, Metaphor and Materiality in Contemporary Cloth

Linda Brassington

BLOOMSBURY VISUAL ARTS
LONDON • NEW YORK • OXFORD • NEW DELHI • SYDNEY

BLOOMSBURY VISUAL ARTS
Bloomsbury Publishing Plc, 50 Bedford Square, London, WC1B 3DP, UK
Bloomsbury Publishing Inc, 1359 Broadway, New York, NY 10018, USA
Bloomsbury Publishing Ireland, 29 Earlsfort Terrace, Dublin 2, D02 AY28, Ireland

First published in Great Britain 2026

Cover design: Holly Capper
Cover image © Linda Brassington (2021). *Compressed.* Textile, paper, foil.
Photo: © Ed Brydon.

A catalogue record for this book is available from the British Library.

A catalog record for this book is available from the Libaray of Congress.

ISBN: HB: 978-1-3504-2386-2
ePDF: 978-1-3504-2387-9
eBook: 978-1-3504-2388-6

Typeset by RefineCatch Limited, Bungay, Suffolk
Printed and bound in Great Britain

For product safety related questions contact productsafety@bloomsbury.com.

This book is dedicated to my mother.

Contents

List of Illustrations		viii
Foreword		xiii
Acknowledgements		xv
Introduction		1
1	The Language of Resist Dyeing	7
	Farnham School of Art: a pedagogy in resist dyeing	11
	Contributions to knowledge: an international perspective	15
	The emergence of a visual and sensory language	18
	Poetic language	26
	The space and place of resist dyeing	28
	Narrative language: emotional connections	30
	The language of intangible cultural heritage	33
2	Movement, Motion and Time: The Formation of Practice	39
	The domestic space: routes and pathways to making	40
	From domestic workshop to professional studio	43
	Material flow: paste and wax	45
	Pattern, rhythm and repeat	47
	Rhythms of resist: 'blueprint'	48
	Rhythms of *shibori*: cultural connections	51
	Rhythms of Mashiko: a personal narrative	55
3	Material as Metaphor: Colour as Stuff	67
	Indigo as substance	69
	The life and death of an indigo vat	73
	Indigo as pigment	76
	Indigo: burnishing and polishing	76
	Colour as stuff	78
	Mud: the passage of time	81
	Minerals and matter as 'place'	83
	Encrusted	88

4 Process as Performance: Repetition, Ritual, Gesture 91
Resist dyeing as bodily action 92
Observers of material flow 97
Performativity as material transformation 99
The subliming vessel 102
Performativity as physical and emotional experience 104

5 Theatres of Making: Space and Place 117
Memories of place 118
Reiko Sudo 122
In search of *ton-byan* 123
Hranovnica: a hiatus in time 125
Púchov, Slovakia 130
Steinberg, Austria 131
Signposts to the future 135
Kyoto, Japan 145
Mashiko, Japan 147
Tokyo, Japan 164
Arimatsu, Japan 167
Mashiko revisited 169
Towards a new definition for intangible cultural heritage 171

List of Artworks 175
References 179
Index 189

Illustrations

1.1 Dress (*c.* 1815–20). Blue, white and chocolate. Vertical stripes of dots cross undulating diagonals. Block printed resist, overdyed in indigo, overprinted with alum, iron and madder from a roller. 13

2.1 Susan Bosence (1983). Resist patterned length. Vat dye over resist paste on indigo dyed cotton, in The Textiles Collection at the University for the Creative Arts at Farnham, 7943. 45

2.2 Linda Brassington (*c.* 1985). Waxed length. Linen, dark blue indigo background with bands of lighter blue and cream spots. Hand drawn wax resist. Spots waxed and overdyed in indigo. Lines waxed and overdyed in myrabolan and iron rust, Crafts Study Centre, 2002.33. 50

2.3 Japanese shibori. Vertical irregular bands of alternate soft blue, dark blue thin moving lines, *c.* early twentieth century. Indigo over hand spun and hand woven cotton from the source collection of Barron and Larcher, The Textiles Collection, University for the Creative Arts at Farnham, 8090. 51

2.4–2.5 The mixing of *katazome* resist paste at *Higeta Aizome Kobo* in Mashiko 2016. 56

2.6 The author applying *katazome* resist paste on linen and *sasawashi* paper cloth at *Higeta Aizome Kobo* in Mashiko 2016. 57

2.7–2.8 Stencils suspended from a drying rack in the wash room at *Higeta Aizome Kobo*, Mashiko, Japan, 2016. 60-61

2.9–2.10 Dyed paper suspended above the indigo vats to oxidize at *Higeta Aizome Kobo*, Mashiko, Japan, 2016. 62-63

2.11–2.14 Movement and motion through the processes of making *Rhythm* in 2017, applying discharge and indigo over masking tape and *katazome* paste resist on linen. 64

2.15 Linda Brassington (2017). *Rhythm* exhibited in *Linda Brassington: Work from the Art Residency 2016* at Mashiko Museum of Ceramic Art, 2017. 65

2.16 Linda Brassington (2017). *Rhythm*. Indigo over discharge and *katazome* paste resist on linen. 65

3.1 Compacted earth surrounding an indigo vat at *Higeta Aizome Kobo* in Mashiko, Japan, 2016. 68
3.2 Anish Kapoor (1992). *Artist's sketchbook.* 80
3.3 Block printed *dabu* mud resist on cloth prior to indigo dyeing, Rajasthan, India, 1989. 82
3.4 Richard Long (2014). *Untitled.* China Clay on plywood. Diptych. 86
3.5 Detail of ochre rubbed over hand-woven ramie and indigo-dyed warp in Chiyoko Tanaka's *'Grinded Fabric – Ocher'* (1991). 87
3.6 Linda Brassington (2016). *Encrusted* (detail). Indigo, pumice and acrylic on Japanese linen cloth. 89
4.1 Detail of Shihoko Fukumoto's indigo-dyed cotton sateen over stitch resist (2001). 94
4.2 Anya Gallaccio (1999). *All the rest is silence* exhibited at Sadler's Wells. Chappa silk, indigo dye, watercolour paper. 98
4.3–4.4 Roger Hiorns (2008). *Seizure.* Copper sulphate crystals. 101
4.5 Installation view of Richard Wilson's *20:50* (1987) at *Space Shifters*, Hayward Gallery, 2018. 106
4.6 Cloth suspended from bamboo stretchers in an indigo vat at *Higeta Aizome Kobo*, Mashiko, Japan, 2016. 108
4.7 Fermenting indigo dye at *Higeta Aizome Kobo* in Mashiko, Japan, 2016. 113
4.8–4.9 Work in progress at *Higeta Aizome Kobo* in Mashiko, Japan, 2016. 114-15
4.10 Indigo vat at *Higeta Aizome Kobo* in Mashiko, Japan, 2016. 116
5.1 Clays extracted from four regions of Kyoto Prefecture in the construction of Hosoo Gallery, Kyoto, Japan, 2019. 121
5.2–5.3 Details of decay in the workshop of Elemír Montško, Hranovnica, Slovakia, 2016. 126
5.4 Interior view of the earth floor at Elemír Montško's workshop, Hranovnica, Slovakia, 2016. 127
5.5 Interior view of the pulley in the dye room of Elemír Montško's workshop, Hranovnica, Slovakia, 2016. 128
5.6–5.7 Collapsed dye vats in the workshop of Elemír Montško at Hranovnica, Slovakia, 2016. 128-29
5.8–5.9 Interior views of the earth floor at Elemír Montško's workshop, Hranovnica, Slovakia, 2016. 129-30
5.10 Helena Trnková and Stanislav Trnka (out of view) pulling out the creases of printed and dyed cloth in Púchov, Slovakia, 1985. 132

5.11 Stanislav Trnka drying cloth from nails in trees at the edge of the riverbank in Púchov, Slovakia, 1985. 132
5.12 Exterior of *Blaudruckerei Koó* at Steinberg, Austria, after snowfall, 2017. 134
5.13 Indigo footprints in snow at *Blaudruckerei Koó*, Steinberg, Austria, 2017. 134
5.14–5.17 Stages of cloth tensioning and indigo dyeing at *Blaudruckerei Koó*, Steinberg, Austria, 2017. 136-37
5.18 Indigo-dyed cloth oxidizing at *Blaudruckerei Koó*, Steinberg, Austria, 2017. 140
5.19–5.20 Blue light installation in *im.material* (Niedermayr, L. and Detschmann, G.) at *Michealertor Kuppel*, Vienna, 2018. 142-43
5.21 Linda Brassington (2018). *Textile installation* in *im.material* (Niedermayr, L. and Detschmann, G.) at *Michealertor,* Vienna, Austria, 2018. 144
5.22–5.23 Linda Brassington (2018). Indigo over Austrian paper cloths produced in collaboration with Josef Koó. 145
5.24 Hiroyuki Shindo (2023). *Shindigo 23–01.* Hand woven cloth (banana and ramie fibre) dyed in Japanese indigo. Exhibited at *Tansa: Japanese threads of influence* (Millar, L.), Crafts Study Centre, 2023. 146
5.25 *Higeta Aizome Kobo* at the crossroads, Mashiko, Japan, 2016. 148
5.26 Interior view of *Higeta Aizome Kobo*, Mashiko, Japan, 2016. 149
5.27 Indigo vats at *Higeta Aizome Kobo*, Mashiko, Japan, 2016. 149
5.28 Gallery talk at the opening of *Linda Brassington: work from the art residency 2016* at Mashiko Museum of Ceramic Art, Mashiko, Japan, April 2017. 153
5.29–5.31 Dyed cloth suspended above the indigo vats to oxidize at *Higeta Aizome Kobo*, Mashiko, Japan, 2016. 154-56
5.32–5.33 The author's *katazome* stencilled cloths suspended from the rafters in preparation for indigo dyeing at *Higeta Aizome Kobo*, Mashiko, Japan, 2016. 157-58
5.34 Linda Brassington (2016). *Diagonal.* Indigo over *katazome* paste resist on discharge printed *sasawashi* paper cloth. 159
5.35 Linda Brassington (2016). *Boundary* (reverse). Indigo over *katazome* paste resist on discharge printed *sasawashi* paper cloth. Collection of Mashiko Museum of Ceramic Art. 160

5.36 Linda Brassington (2016). *Ellipse 2.* Indigo over *katazome* paste resist on discharge printed *sasawashi* paper cloth. 161
5.37–5.38 Gallery views of the exhibition *Linda Brassington: work from the art residency 2016* at Mashiko Museum of Ceramic Art, Mashiko, Japan, April–June 2017. 162
5.39 Linda Brassington (2016). *Pierced.* Indigo over *katazome* paste resist on discharge printed *sasawashi* paper cloth. Collection of Mashiko Museum of Ceramic Art. 163
5.40–5.41 Finished garment at Homme Plissé Issey Miyake, Tokyo, Japan, 2019. 166-67
5.42 Linda Brassington (2021). *Compressed.* Textile, paper, foil. 168
5.43 Linda Brassington (2023). Painted discharge paste on indigo-dyed cloth at *Higeta Aizome Kobo*, Mashiko, Japan, 2023. 171
5.44 Linda Brassington (2023). *Katazome* pasted cloth suspended on bamboo stretchers in indigo at *Higeta Aizome Kobe*, Mashiko, Japan. 2023. 172
5.45 Linda Brassington (2023). Indigo dyed hemp fibre at *Higeta Aizome Kobo*, Mashiko, Japan, 2023. 173
5.46 Linda Brassington (2023). *Entangled.* Ramie in indigo at *Higeta Aizome Kobo*, Mashiko, Japan, 2023. 174

Foreword

by Lesley Millar
Emerita Professor of Textile Culture, University for the Creative Arts

In the age of the internet and social media there has emerged an expectation that everything can be clarified and verified. There is also, however, a deep hunger for mystery. Our knowledge of indigo firmly places it within a botanical context, satisfying our first criterion for factual evidence, and, wonderfully, it also satisfies our second requirement for an ambiguity seemingly beyond our control. Dyeing with indigo is the nearest we come to pure alchemy, as it transitions, before our eyes, from one state to another.

Indigo means so many different things. It is a plant, a dye, a colour, an identity, and even, as the great jazz musician Duke Ellington proposed, a mood. And he was not wrong – working with indigo is an experience, a performance and a relationship. We must be sensitive to the substance as we gently lower the material beneath its crust, disturbing the liquid as little as possible while ensuring the movement essential to the penetration of the dye. There is the anticipation as the material and the dye begin to work together, hidden from view in the depths of the of the dye liquid. This is followed by a sense of anxiety when we remove the material from the dye vat, and then the sheer delight as the colour oxidizes and becomes that unique blue we know as 'indigo'. It is a blue that can range from the palest sky blue all the way to the deepest blue/black.

Linda Brassington, through her extensive knowledge, skill and understanding, acquired over many years, describes the immersive qualities of indigo particularly as a means of interrogating our relationship with space and place. There have been scholarly texts published about the history and tradition of indigo, its indigenous roots and the trade routes carrying the plant and the dyed materials across the world. There are also many books giving practical advice on how to use indigo as a dyestuff. However, no one, until now, has provided a text that gives the reader an insight into the poetic, spatial and experiential qualities of indigo that can act as a gateway to all of the above.

Drawing on the writing and practice of those outside the field of textiles, for example cultural theorist Homi K. Bhabha and social anthropologist Tim Ingold, and from textile practitioner diaries, including her own, she establishes the case for indigo to be accepted as Intangible Cultural Heritage. In this book, Linda Brassington has developed her ideas *about* indigo, and her work *with* indigo, to evidence that it is an appropriate medium for contemporary art concerned with spatial practice, one that is innovative and grounded in tradition. Most importantly, she reveals the potential for indigo to be the locus of our wish to experience a mysterious transformation, and our sensual appreciation of that deep, deep blue.

April 2024

Acknowledgements

I am indebted to Lesley Millar and the late Simon Olding for their guidance and encouragement in the development of my research from inception to completion.

My research has developed through numerous field study visits, working collaborations, exhibitions and art residencies in Central Europe and Japan. Special thanks are given to the many friends, colleagues and institutions who have supported me on these journeys.

Early research in former Czechoslovakia was supported by the Churchill Trust and the British Council. That research would not have been possible without the support of Zdena Taptuch. I am grateful to her family for their hospitality and for my early introduction to 'blueprint'. I am indebted to Svätopluk Hlavačka for accompanying me on my field study visits in Slovakia, for his enduring patience and continuing support for my research. I am grateful to the late Helena Trnková and Stanislav Trnka for their warm welcome at their workshop during my early visits to Púchov. I thank Martin Trnka for arranging my subsequent visit to his father's workshop in 2017, and Peter Trnka for his guidance in the application of the resist paste 'pap'. Katarína Podobová was instrumental in introducing me to ÚĽUV, the Centre for Folk Art Production in Slovakia, and I am grateful to ÚĽUV for inviting me to contribute to their colloquium and exhibition *Modrotlač* in Bratislava in 2014, and for the publication of my research in their journal and at their symposium in Košice in 2016. I thank Juraj Podoba for his introduction to the workshop of Elemír Montško in Hranovnica, and for sharing his ethnographic research with me (Podoba 1988). I am grateful to Elena Bekešová at the Podtatranské Museum in Poprad for arranging my subsequent visit to the Montško workshop in 2016, and to the community of Hranovnica for enabling my research.

In Japan, I am indebted to the Mashiko Museum of Ceramic Art for my residency at *Higeta Aizome Kobo* in 2016, for their exhibition of my work in 2017 and for my subsequent visit in 2023. Special thanks go to Tadashi Higeta for the opportunity to develop my practice in the working environment of his unique heritage site. I wish to thank curators Yuko Matsuzaki and Satoshi Yokobori, the residency programme organizers Mayumi Tsukimura and Masako Onishi, interpreter Reiko Matsuzaki and all the staff at the museum for their support. I am particularly grateful to Takumi Kojima and all the dyers and craftspeople at the *Higeta Aizome Kobo* for their generosity, valuable knowledge and shared experience. My thanks to Keiko Kawashima and Reiko Sudo for enabling research visits in Japan in 2019 to Edo Komon in Tokyo, Hiroyuki Shindo, Chiyoko Tanaka and Jun Tomita in Kyoto, Harisho, Kunosenkoujo and Suzusan in Arimatsu, and Kaga-Yuzen in Kanazawa for sharing the specialist practices that have informed my research. I am grateful to the Fantasia film production company, director Shigeru Yoshida and translator Riki Ohkanda for their invitation to contribute to the documentary *Blue Comes from Indigo, Yet it is Bluer than Indigo Itself* (2024).

In Austria, thanks are due to curators Lisa Niedermayr and Gabriele Detschmann for their exhibition of my work in *im.material* in St Michael's Tower at the Hofburg Palace in Vienna in 2018. I am grateful to them for arranging my collaboration with *Original Blaudruckerei Koó* in Steinberg, and to Josef Koó for his shared knowledge and support in the realization of the project. I thank UNESCO Austria for inviting me to contribute to their dialogue with resist dyers from Austria, the Czech Republic, Germany, Hungary and Slovakia in Gutau in 2016, and for their consultation in the nomination of resist block printing and indigo dyeing in Europe to the Representative List of the Intangible Cultural Heritage of Humanity in 2018.

I am grateful to the Great Britain Sasakawa Foundation, the Worshipful Company of Dyers, the Society of Dyers and Colourists, the Pasold Research Fund and the University for the Creative Arts Research Fund for their financial support. Furthermore, my practice and research has been supported by my colleagues in the School of Textiles, the Crafts Study Centre, the International Textile Research Centre, the Visual Arts Data Service and the Research Office at the University for the Creative Arts. Warmest gratitude is due to all of them.

Finally, I am especially grateful to my husband, daughter and all my family and friends for their steadfast support and encouragement throughout every stage of my research. This outcome of my research is dedicated to my late mother.

Introduction

This book introduces new interpretations of indigo and resist dyeing through the immersive experience of dye and cloth. Making is at the heart of its exploration, generating new meaning in the vacant spaces between historical, ethnographic and technical perspectives on the subject. The author's reflective practice is a fundamental element of the book. It is an account of indigo and resist dyeing through art residency, working collaboration and field study in Central Europe and Japan – locations that have been the subject of experiential research since 1985. As the exploration evolves, a choreography of craft is recorded in 'theatres of making' in which field study and textile practice become a conduit for the transmission of craft heritage, and the final chapter develops towards a new definition of intangible cultural heritage in resist dyeing.

The author's work is developed against a backdrop of learning through practice, formerly as a student of textiles and currently as an artist and teacher. The research focuses on sources that have informed their knowledge and appreciation of resist dyeing, initially in the context of art education and subsequently in the development of contemporary cloth. The book begins with a review of 'The Language of Resist Dyeing' revealed in historical literature and more recent publications; identifying the approaches taken by other scholars and practitioners. It is followed by a chapter on the 'Movement, Motion and Time' of resist dyeing, the rhythms of repeat that are central to the field of printed and dyed textiles and the routes that are mapped through its workspaces. The chapter highlights important figures of the twentieth century who have adopted and developed indigo and resist dyeing in modern textiles, paving the way for new interpretations in contemporary practice.

In Amsterdam in 2009, attendance at the conference *Showing Making* set in motion a period of consideration as to how the distinctive materiality and making in resist dyeing might be shown. The anthropologist Tim Ingold introduced his definition of making as 'the intervention between fields of force and flows of material' (2009a). This assertion has been key in guiding the author's direction; asking what are the 'fields of force and flows of material' in resist dyeing, and how are they defined in contemporary textile practice?

Indigo is commonly, though not exclusively, associated with resist dyeing. As the least penetrating of dyes, indigo offers the greatest scope for clarity in resist dyed mark making. Significantly, as a plant dye extracted from *indigofera tinctoria*, dried indigo does not dissolve naturally in water. Instead, with the aid of additional chemicals, the dye is drawn down into the vat, oxygen is raised to the surface in an accumulation of bubbles, and the indigo undergoes a process of reduction that results in a lime green liquor beneath a surface of fermenting froth. The distinctive colour associated with indigo occurs through a process of oxidation, a development from yellow to green to blue as the cloth emerges from the dye vat and is exposed to air. Indigo has an ancient history that precedes modern printed textile practices and is evident in many world cultures. Consequently, its practices have given rise to a breadth of historical research, technical analysis and ethnographic surveys. Indigo has unique material qualities. It has a propensity not only to change colour (through oxidation) but also to change its surface quality. It is a substance with the capacity to act both as dye and pigment: through the cloth and on its surface. The chapter 'Material as Metaphor' demonstrates the power of indigo to express volume, an expanse of space, an all-enveloping darkness; and illustrates mud as an iron-rich colour and as a veil on the surface of the earth that moves, changes and shifts. The chapter draws attention to the important research undertaken by other scholars of indigo, mud and tannins, and draws comparisons with artists in other disciplines whose works exploit the stuff of colour and the performativity of material, their connection to the earth and the ground beneath our feet.

Resist dyeing is pattern making in reverse. Its practices include a range of techniques from the application of resistant substances – paste, wax, resin – to the manipulations of material – folding, wrapping, binding, stitching – to prevent the penetration of dye into cloth. Its results are dependent not only on the techniques applied, but also on the variations of woven structure and cloth substance in the materials employed. Any play or movement in the fastenings results in the bleeding of dye into fibre, the penetration of colour into thread, the seepage of stains into cloth. Marks on manipulated cloth are the result of physical force, the indirect result of actions on cloth. Their processes are charged with energy – a vigorous tightening of knots, the dynamic rotation of wrapped threads around cylindrical forms; even the smallest of stitches requiring a stabbing of the needle through cloth. As such, bodily engagement is intrinsically bound within resist dyeing. The chapter 'Process as Performance' explores the actions of resist dyeing and the emotional resonance they evoke. It draws parallels with artists' works in which the performativity of its processes triggers acute reactions – either in its material effects or in the response of its audience.

The overarching question – what are the 'fields of force and flows of material' in resist dyeing, and how are they defined in contemporary textile practice? – is examined through the theories of three key thinkers: Ingold, cultural theorist Homi K. Bhabha and geographer Yi-Fu Tuan. The author draws upon Ingold's ideas on the relationship between environmental perception and embodied skill, and questions how the dynamics of movement impact on creative practice in resist dyeing. In parallel, Bhabha's expressions of performativity in materials and processes are explored through his critiques of art practice, questioning how they might be reinterpreted as metaphorical meaning in resist dyeing. As the study moves to sites of practice in England, Central Europe and Japan, the relationship between resist dyeing and its working environments are explored in the context of Tuan's definition of space and place. The author considers the significance of physical engagement in heritage sites, and examines the influence of practice in the making of place. Together, the selected theories of Ingold, Bhabha and Tuan inform a collection of ideas and concepts that contrast and converge within and across the chapters to form the unfolding shape of the book.

The relationship between intangible cultural heritage and the development of contemporary practice is an emerging theme and culminates in the chapter 'Theatres of Making'. The author's international field work, art residencies and exhibitions open up an experiential space through which to develop and articulate ideas. While the chapter reflects on UNESCO's objectives for the safeguarding of traditional craft (2003a) – preserving cultural heritage, respecting communities, educating people, sharing knowledge, raising awareness, and so on – it presents an account of how individual practitioners, in their unique sites of indigo dyeing and resist, hold the materiality of their heritage; how it is demonstrated in their sensibility and in their knowledge and understanding of process. This chapter presents artists and artisans as the personification of intangible cultural heritage embodied in their respective practices, and introduces the proposition that intangible cultural heritage is communicated in the sensory experience of their workspaces.

Buildings are essential in holding the memory of indigo within their spaces. To separate the processes and practices of resist dyeing from their physical environment would be to diminish their meaning and authenticity. Instead, this book explores contrasting sites, and how the materiality of resist dyeing is revealed in the very fabric of the buildings they occupy. In parallel, the significance of international collaboration and exchange in Central Europe and Japan, the lines of communication they have created and the network of ideas they have generated, express Bhabha's notion of heritage as a 'creative opportunity' (*Discussing the Potentials of Cultural Heritage in Austria: Press Conference with Homi K. Bhabha*, 2006). The author's direct experience of workspaces comes to redefine intangible cultural heritage. In these theatres of making the intangible is made tangible in another form, the reconfiguration of traditional craft in contemporary practice.

This book sets out a new space for an appreciation of intangible cultural heritage. As the author traces pathways through dye workshops past and present, the sense of memory and loss held in the spaces marks out their place in the embodiment of intangible cultural heritage. Language in the imagining and creation of place – the communication of sensory spaces in indigo and resist dyeing – becomes increasingly important. With this objective in mind, the author applies two contrasting writing styles to emphasize two parallel voices: theoretical and lyrical. This could be described as the analytical and objective alongside the performative and subjective. Lyrical accounts, signified by passages in *italic*, denote the experiential voice interwoven through the book, and an understanding of indigo and resist dyeing is expressed in the resonance of poetic language – both linguistically and through physical and creative reimaginings in cloth. They express the rhythm of making in contrasting sites of practice, questioning how the processes and their associated substances and dyestuffs reveal new meaning. In doing so, the author proposes a new way of writing about these materials and practices to develop a fresh understanding of the subject. As such, the spirit of resist dyeing is inextricably linked to cloth. The narratives reveal the dark spaces of indigo dyed textiles, the layers beneath the heavily pigmented surfaces of textile works. The material substance of indigo unfolds to reveal new meaning in the history and culture of resist dyeing, and its intangibility is manifested in the space and place of the subject. Tuan observes:

> We may speak of a life of feeling as we do of a life of thought. [But] it is a common tendency to regard feeling and thought as opposed, the one registering subjective states, the other reporting on objective reality. In fact, they lie near the two ends of an experiential continuum, and both are ways of knowing.
>
> Tuan 1977: 10

In this book, practice combines with sensibility to contribute a philosophical and conceptual perspective to the breadth of historical and technical knowledge of indigo. Resist dyeing is defined as a performative process, and establishes a deeper, metaphorical interpretation of its materiality. The author's experiential research marks a transition between traditional craft and current practices – the then and now. It signals new meaning in indigo and resist dyeing, and identifies the role of contemporary practice in defining intangible cultural heritage in the subject. Through the exploration of metaphorical and performance meanings in resist dyeing, this work moves towards a new critique for the classification of Intangible Cultural Heritage. It stretches the boundaries of UNESCO's identification, redefining it beyond skills and knowledge and into a sensory space of lived experience.

1

The Language of Resist Dyeing

The history and technique of resist dyeing are embedded in its written language, and their details form the majority of its literature. This book presents a new reading of textile text where nuances of description and expression in resist dyeing are put forward. It applies textile thinking to this process – to unlock its complex meanings and to establish a new approach to writing about the subject.

This chapter identifies the vacant spaces in the language of resist dyeing where metaphor and performance can contribute new meaning to the subject. The starting point is a review of the written language of indigo and resist dyed textiles that has gone before; to examine how the history and techniques of indigo have been described, how the actions employed in resist dyeing have been discussed. The review questions the apparent absence of physical and emotional encounter in the subject, and considers how perceptions of its cloth and the character of its material substance have been expressed. The chapter explores the visual and sensory language that surrounds indigo and resist dyeing, how language – narrative, poetic, lyrical – has been used to capture these qualities. It discusses a range of personal voices in literature that have begun to articulate the characteristics of resist dyeing, celebrating its many variables and conveying its energy.

Indigo is the substance of resist dyeing and is deeply rooted in its history and technique. From the thirteenth century onwards, as indigo passed through North Africa into European culture, its rediscovery became a source of scientific wonder. But, as industrial and technological advancements progressed, indigo came under threat. Driven by a compulsion to capture knowledge before it was lost, scholars embarked on a process of preservation through technical records. The result was an emphasis on factual information and a detachment from the sensory interaction with indigo and resist dyeing. Texts shifted away from resonance and reimagining towards restoration and remembrance.

In 1813, in his *Philosophy of Permanent Colours*, Fellow of the Royal Society of London Edward Bancroft had introduced indigo as 'the most interesting, important, and instructive [subject] which can occupy the attention of a dyer or chemist' (Bancroft 1813: 165). Describing the properties of indigo as 'admirable and singular', their constituent effects became the subject of Bancroft's intense research and experimentation for almost twenty years.

In 1856 British chemist William Perkin discovered aniline dye, the forerunner of synthetic colour that was to transform the global production of textiles. But in *Mauve: How One Man Invented a Colour that Changed the World* (Garfield 2000), Simon Garfield writes: 'Throughout [Perkin's] industrial career, one colour eluded him and his colleagues, and remained the Holy Grail of the synthetic dye trade until the end of the century: indigo' (Garfield 2000: 124). By 1890 a synthesis of indigo was developed by BASF in Germany. 'Within two decades it had devastated the traditional supplies from India' (Garfield 2000: 126). Indigo had dominated the application of blue on cloth from antiquity to the nineteenth century, but in the Preface to his research Bancroft stressed:

> I have endeavoured to avoid minute descriptions of such manual and mechanical operations in dyeing and calico printing in common use, and not connected with the *principles* of the art . . . The artists know these operations already, and to mere speculative or philosophical readers, a knowledge of them would be useless.
>
> Bancroft 1813: vii

It is discernible, therefore, that during the nineteenth century the importance of indigo was universally understood, while an acknowledgement of its detailed practices and their significance was less apparent.

In 1969 Stuart Robinson wrote *A History of Printed Textiles*. The publishers Studio Vista London wrote:

> This is the first book in English to explore the fascinating story of man's attempts to print patterns on cloth. Previously the information has been available only in various articles, journals and books, which are mainly in foreign languages or long out of print.
>
> Robinson 1969a: 1

In the companion volume, *A History of Dyed Textiles*, Robinson wrote:

> Despite the conditions reflected in an early papyrus which stated that the 'hands of the dyer reek like rotting fish and his eyes are overcome with weariness', the development of dyeing techniques persisted from earliest known civilisations to the present day. As man discovered new natural fibres, and later synthetic fibres, so he invented new dyes and dyeing methods.
>
> Robinson 1969b: 1

Robinson's two companion publications were intended as 'a comprehensive source book for students preparing special studies upon printed textiles' (Robinson 1969a: 1). They included references 'upon textile dyeing, including tie-dye and batik methods' that were otherwise 'widely scattered' (Robinson 1969b: 1). Significantly, the limited penetration of indigo into natural fibres – cotton, linen, silk – presented considerable potential for creative mark making in resist dyed textiles.

Robinson grouped resist processes into the categories of batik (protection of the cloth from the dye by the drawing or printing of wax), 'tie and dye' (where parts of the cloth are tied up, knotted or stitched and the threads drawn up before dyeing), clamping (where areas of the cloth are compressed between plates) and reserve patterning (where a pattern is painted, printed or stencilled onto the cloth with a resist paste – rice, flour, mud, clay or resin). Robinson also refers to *ikat*. Known as *kasuri* in Japan, *ikat* is a variation on tying, but in this technique, ties are applied to hanks of yarn to resist the threads from dye prior to weaving (Robinson 1969b: 39). However, as this book focuses specifically on resist dyeing on cloth, *ikat* is not discussed further in this review.

Robinson's work marked a turning point in the specialist study of printed and dyed textiles during the second half of the twentieth century. While his research is described as 'a history', it presented an enlightened understanding of history, not only providing comprehensive information on scientific developments including classifications of dyes and fibres, but also contributing detailed descriptions of the breadth of techniques in resist processes worldwide. Each volume includes appendices with extensive lists of textual references, international museums and collections, contributing valuable educational resources for study in the subject.

Robinson stresses the significance of indigo in the context of world history. Indigo was one of the most widespread and important of early dyestuffs. It spanned the whole of Asia, large areas of the African continent, across the north, south and central Americas and the regions of New Guinea. Indigo became the universal dye for blue and formed the base colour for subsequent overdyeing with other natural extracts to achieve secondary and tertiary colours. By the sixteenth century, indigo and chintz were as important as spices in the development of East India trading. Despite its high price, indigo, known as 'blue gold' to reflect its significant value, soon replaced European woad for blue, and its cultivation in India became a major industry of the British Empire. In a review of the indigo export trade in 1912, the Royal Economic Society recorded the export of indigo from India between 1895 and 1896 as 9,366 tons valued at

£3,566,700. By 1910 to 1911 export declined to 846 tons valued at £223,500 (Royal Economic Society 1912). During that fifteen-year period, large-scale production on plantations in the West Indies had brought down the price, while, in 1897, the synthesis of indigo led to the production of a laboratory-produced dyestuff. Robinson explains that in 1834 Friedlieb Ferdinand Runge produced a new kind of dye called kyanol that started a gradual process of experimentation in synthetic colourings until William Perkin's discovery of mauveine in 1856 (Robinson 1969a: 23). Eventually, natural indigo was 'defeated by the rapid advance of the coal tar industry that grew up in the second half of the nineteenth century' (Robinson 1971: 23), and by 1913 synthetic indigo had almost completely superseded the natural product.

Prior to Robinson's publications, Phyllis Barron and Dorothy Larcher were leading studio practitioners in the development of modern printed and dyed textiles in England during the first half of the twentieth century, and they continue to inform this book. Many examples of their own source collection of textiles demonstrate the qualities of indigo and resist dyeing. Barron and Larcher's source collection was the focus of earlier research in *The Collection as a Narrative in the Crafts* (Brassington 2004), previously published by the Crafts Study Centre specialist university museum, Farnham, as online learning material. The collection is discussed again in the chapter 'The Muriel Rose Bequest' (Brassington 2006: 49) in *Muriel Rose: A Modern Crafts Legacy* (Vacher ed.), also published by the Crafts Study Centre.

Research towards both publications through *The Textiles Collection*, previously held at the University for the Creative Arts at Farnham and currently at UCA Epsom, revealed a clear connection between Barron and Larcher's own collection of world textiles and their practitioners' understanding of resist dyeing. They include an Indian muslin, finely drawn with molten wax to resemble a print in repeat, and subsequently overdyed in madder and indigo (ID Number Current Accession 8045). And there is a large double-width cotton piece originating from West Africa, tightly bound with close clusters of spots. It is still tied with fine raffia and caked in the residue of unwashed indigo left behind from its making (ID Number Current Accession 8052). It is probable that Larcher brought back some textiles from her travels in India, while others were most likely gathered from English and French markets or collected from junk shops. Having been previously discarded or sold on, these textiles were rediscovered by Barron and Larcher and, together, became reference points for their work.

Barron's research into the application of natural dyes on cloth, principally cottons and linens, gave rise to a distinctive palette in modern block printed textiles which she created with Larcher. In post-Second World War art education, Barron's knowledge and understanding came to underpin crafts education at Dartington College in Devon where Robin Tanner, Her Majesty's Inspector (HMI) in primary education, provided a series of in-service Ministry of Education training events for teaching professionals across all sectors of education. To illustrate his lectures, Tanner mounted a series of temporary displays against rolls of undulating corrugated card. His idiosyncratic compositions incorporated objects of historical, visual and material significance. They often included examples of work by Barron and other artists, demonstrating Tanner's value for craft and his passion for authenticity in design and making. Later Tanner became a founder trustee of the Crafts Study Centre. The bequest of the Barron and Larcher textiles to Tanner in 1964, and his response to the body of work entrusted to his charge, impelled him and others such as Muriel Rose to initiate the Crafts Study Centre project. Eventually, a trust was established in 1970 at the Holburne Museum, Bath, moving to the University for the Creative Arts, Farnham, in 2000.

Farnham School of Art: a pedagogy in resist dyeing

Farnham School of Art was founded in 1866. In 1969 it became the West Surrey College of Art and Design, and the University for the Creative Arts since 2008. During the 1950s and 60s, Ella McLeod, founder of the Textiles course at Farnham School of Art, engendered a deep philosophy of woven textiles through an acute understanding of materials and processes. When print was introduced, McLeod's pedagogy was further shaped by Deryn O'Connor as principal lecturer in printed and dyed textiles.

In the 1960s and 70s, in parallel with Robinson's informative and instructive publications, resist dyeing emerged as an element of textile studies in twentieth-century art education. Embedded within the printed textiles curriculum, but with its roots in textile tradition, resist dyeing was somewhat overshadowed by the more modern methods of printing with block and screen. In a few art institutions, such as the West Surrey College of Art and Design, pedagogy embraced the potential of resist dyeing as an activity for imagination and creative expression, sometimes combined with other materials and contemporary processes. During the 1970s and 80s O'Connor developed a close affinity with Susan Bosence, a block printer and resist dyer who had been mentored by Barron during her early practice. O'Connor and Bosence promoted and developed the study of resist dyeing at Farnham, while Bosence applied similar principles with students at Camberwell School of Arts and Crafts and the Adult Education Centre, Dartington. This relationship was significant in the dissemination of their knowledge. In 1985 Bosence's *Hand Block Printing and Resist Dyeing* was published by David and Charles. In Bosence's consideration of patternmaking and designing she wrote:

> Suppose one is interested in lines . . . Designs of plain stripes can be very elegant and satisfying if the method of patterning is appropriate . . . repeating a plain line with a block may not be right, as the joins are very difficult to mask. This is why some of the most satisfactory linear designs are made by the roller printing machine, which prints a continuous line from an engraved roller automatically charged with dye as it rotates. Linear prints by block can be interesting if the 'join' is intentionally designed . . . For my part I am happy to work in a resist method . . . paste, stitch or wax to produce lines whose character and texture evolve from the way they are made.
>
> Bosence 1985: 89–90

In 1982, at the James Hockey Gallery at the West Surrey College of Art and Design in Farnham, O'Connor with Hero Granger-Taylor curated the exhibition *Colour and the Calico Printer: An Exhibition of Printed and Dyed Textiles 1750–1850* (Granger-Taylor and O'Connor 1982) with the aim of exploring the use of dyes on printed textiles before the introduction of synthetic dyes in 1856. Exhibits were grouped according to colour. In the accompanying catalogue, the section on 'Blue' distinguished the treatment of indigo as distinctly different from other colours in calico printing. It described the complexities of the vat dyeing process, and explained the methods of resist printing employed to reserve areas of pattern prior to dyeing. Granger-Taylor and O'Connor went on to describe the subsequent technical explorations in the complex and precarious attempts to harness indigo as a print medium rather than as a resist process.

Granger-Taylor and O'Connor stressed that their research was carried out within the 'limitations' of any prior training in applied chemistry but, because all of their analysis had been 'by visual means only', they felt that 'the information given in the actual catalogue entries embodies our contribution to the subject' (1982: 8). The detail of each listing is testimony to Granger-Taylor and O'Connor's shrewd observation and their detailed scrutiny of each carefully selected example. In a catalogue entry relating to a blue, white and chocolate-coloured dress of 1815 to 1820, Granger-Taylor and O'Connor describe the complexities of an otherwise superficially simple design in which 'vertical stripes of dots cross

Figure 1.1 Dress (*c.* 1815–20). Blue, white and chocolate. Vertical stripes of dots cross undulating diagonals. Block printed resist, overdyed in indigo, overprinted with alum, iron and madder from a roller. Devonshire Collection of Period Costume, Totnes. Reproduced with the permission of the University for the Creative Arts.

undulating diagonals'. The cloth has been block printed with a resist paste and overdyed in indigo, and the resultant waves of pale blue and white have been subsequently overprinted with a roller of dark dotted stripes. Through careful analysis of the print, Granger-Taylor and O'Connor deduced the measurement of the block and the repeat of the roller: 'It is possible to see in the vertical stripes of dots a regular mark 11.8cms apart. This represents the circumference of the mill and denotes the edge of the pattern on the die' (1982: 20) – 'die' referring to the block or stamp. Granger-Taylor and O'Connor's observations provide further insight. Importantly, they draw attention to this 'charming dress' as an example of how designers have applied visual judgement and creativity within the limitations of a colour that can be dyed but not printed: 'The proportion between blue and white is about equal to make it appear almost a positively printed colour' (1982: 17). Significantly, Granger-Taylor and O'Connor point out extraordinary examples of resist-dyed calicoes originating from India where molten wax has been applied, often over the whole of the cloth, to confine indigo blue within very small spaces and narrow channels of the design (1982: 58).

In her 1985 Foreword to Bosence's *Hand Block Printing and Resist Dyeing*, O'Connor wrote: 'One main message comes through . . . the use of dyeing techniques helps to develop an awareness of the cloth as a whole' (1985a). Clearly, resist dyeing played a central role in their shared pedagogy of printed and dyed textiles. Bosence's book was important in documenting a history of the subject. It offered practical explanation and guidance in its techniques while sharing Bosence's personal responses to a breadth of references – publications, artefacts, archival material – and discussing their relevance from the perspective of a practising artist. O'Connor's insightful Foreword presents a valuable understanding of Bosence's practice and her contribution to resist dyeing:

> This book is a rare event . . . Susan Bosence shares with us the steps in the development of her own work . . . It is not a blueprint for the creation of textiles. It cannot be or it would be suffocating . . . As a teacher [Bosence] has always encouraged students to explore something outside themselves and then to transmute this into cloth form. It is important to communicate something, not to experiment for experiment's sake.
>
> O'Connor in Bosence 1985a: 7–8

Following her death in 2018, O'Connor's extensive collection of textiles revealed numerous samples of her own, together with test lengths of resist-dyed pieces recording the creative possibilities of a technique – a row of stitches, a cluster of dots, a scattering of spots. Their details have been discovered in a simple cardboard folder of O'Connor's handwritten technical notes. Such was her adherence to record-keeping that each sheet was accompanied by a diagram to illustrate her stitching plan or composition of waxed markings. These explorations on cloth reflect O'Connor's personal commitment to continued practice and learning – her resolve to understand methods and materials first hand – and they have been bequeathed to the Textiles Collection at the University for the Creative Arts for future study.

Contributions to knowledge: an international perspective

Two international publications in the English language have been significant in the study of resist dyeing – *Indigo Blue Print in Slovak Folk Art* by Josef Vydra (1954) and *Shibori: The Inventive Art of Japanese Shaped Resist Dyeing* by Yoshiko Wada, Mary Kellogg Rice and Jane Barton (1985).

In parallel with Tanner in England, Vydra was a teacher of art education in former Czechoslovakia and a prominent art historian and ethnographer. He was delegated to contribute Czechoslovak pedagogy at international events including in England, and he represented the country on international committees for art education and applied arts. Founder of the School of Applied Arts in Bratislava, Vydra was regarded as a pioneer in art theory, particularly in the fields of industry and folk art. Between 1923 and 1950, he researched and photographed the 'then-alive traditions, various working acts, and folk makers at work in Slovakia' (Čechmánková 2012). *Indigo Blue Print in Slovak Folk Art* was published by Artia Prague shortly before Vydra's death in 1956 and was translated from Slovak into English and French. It was the first publication dedicated to documenting the history, process and traditions of resist printing and indigo dyeing that, in their various forms, extended across mainland Europe at that time and appeared in workwear and traditional costume throughout the region.

Vydra's publication formed the basis for earlier fieldwork in former Czechoslovakia that resulted in the exhibition and accompanying catalogue *Modrotlač: Indigo Country Cloths and Artefacts* (Brassington 1987) published by the West Surrey College of Art and Design. In 2014 the Centre for Folk Art Production (ÚĽUV) published new research by Oľga Danglová in *Blueprint in Slovakia*. In 2017, in their nomination and accompanying online video titled *Blaudruck/Modrotisk/Kékfestés/Modrotlač: Resist Block-printing and Indigo-dyeing in Europe*, the Austrian Commission for UNESCO highlighted the practices of the craft across Central Europe, and in 2018 'blueprint' in the five nations of Austria, the Czech Republic, Germany, Hungary and Slovakia was added to the UNESCO Representative List of Intangible Cultural Heritage.

In 1985 Kodansha America published the first comprehensive work on *Shibori: The Inventive Art of Japanese Shaped Resist Dyeing*. In defining *shibori*, the authors Wada, Kellogg Rice and Barton used the word *shibori* 'because there is no English equivalent'. Importantly, the term encompassed all the various techniques and methods that previously had been incorrectly 'lumped together' as 'tie-dye'. In their view, the word *shibori* seemed 'the most useful term for the entire group of shaped resist textiles', and they hoped that the term would 'win acceptance in the international textile vocabulary' (1985: 7). The publication, rich with explanation, detailed diagrams, photographs and illustrations, is the seminal reference book on *shibori* in the English language. In particular, the publication supports the practical application of *shibori* techniques. Its visual content underpins creative investigations on cloth while contextualizing the craft in the history of Japan. It outlines developments from ancient burial sites, where fragments of resist-dyed cloths were first discovered, through to the period when Japanese society was closed from Western interaction. It explains how, after the Second World War, the country opened up and *shibori* evolved from tradition into contemporary practice.

With the onset of the new millennium, the works of Wada et al. and Vydra have become key texts in the understanding and application of resist dyeing in the second half of the twentieth century and they remain fundamental in informing this book. During the same period other scholars, most notably in museums, contributed a range of academic research on aspects of resist dyeing, its history and practices. In Washington in 1982, the Textile Museum staged the exhibition *Master Dyers of the World* and published an accompanying catalogue by Mattiebelle Gittinger, whose substantial research used George Hewitt Myer's collection of Indian resist-dyed textiles as her 'nucleus' (1982: 7).

Prior to her appointment as the inaugural Thomas Jaffe Curator of Indo-Pacific Art at Yale University in 2009, Ruth Barnes' research into Indian textiles in the Ashmolean Museum, Oxford, contributed substantial knowledge in the field. In her 'Introduction to the Percy Edward Newberry Collection at the University of Oxford', Barnes defines the catalogue of over 1,200 examples as 'outstanding in size' and 'remarkable for its variety', exceeding the Victoria and Albert Museum, London, the Calico Museum in Ahmedabad, India and other notable international collections (Barnes 1997a). Her in-depth analysis at the Ashmolean, together with her survey of South East Asian textiles in European collections, has established Barnes' authority in the subject. The collection at the former *Museum für Völkerkunde* (now the Museum of Cultures) in Basel, Switzerland, holds an outstanding body of indigo and mineral dyed textiles from Asia that illustrates the richly polished indigo surfaces developed in that region, and demonstrates the breadth of resist-dyed textiles from China, Japan and South East Asia. In her research on 'Resist- and mordant-dyeing' Barnes acknowledges the contribution made by the museum's curator Alfred Bühler in his definition of different forms of resist dyeing. In Barnes' opinion, Bühler has 'done much to further our understanding of the history and production of fabrics' (Barnes 1997b).

Research on early explorers and ethnographers by distinguished academics, and the significance of the artefacts they brought back, illustrates the extent to which Indian and South East Asian textiles have informed and inspired the development of printed textile production in England and Western Europe in the eighteenth and nineteenth centuries. While the large-scale application of natural dyes declined, research analysing the complexity of their processes and patterning procedures continued to underpin the development of printed textile manufacturing into the twentieth and twenty-first centuries.

Subsequent to her *A Manual of Textile Printing* (Storey 1974), published by Thames and Hudson, Joyce Storey, senior lecturer at Manchester Polytechnic (now Manchester Metropolitan University), with her extensive knowledge of textile manufacturing in Lancashire, published *A Manual of Dyes and Fabrics* (1978). It included a detailed technical explanation of so-called 'African printing', a term for the popular 'West African Prints' produced in Manchester by wax and chemical resist print methods on cotton for markets in Nigeria and neighbouring West African countries. Storey's work illustrates the irony of how a declining craft of batik in South East Asia has gone on to stimulate alternative industrial processes in Western Europe that have resulted in new markets in Africa. As such, 'West African Prints' are a narrative in cultural exchange, the transfer of textile traditions between nations and across commercial markets. While these developments may be to the detriment of some and to the advancement of others, batik has been transposed into another context, and now these prints have come to symbolize the traditional dress of many West African countries.

In 1989 ethnographers John Picton and John Mack contributed their extensive research in *African Textiles: Looms, Weaving and Design*, published by the British Museum Press. It was the first ethnographic survey of textiles from the continent of Africa. 'Pattern dyeing', the single chapter that focused on resist dyeing, identified regional differences in resist techniques, and offered a brief but significant description of 'mud cloth'. The *bogolan* cloths of Mali are developed on the principle of printing a mud paste, containing elements of iron, onto a cloth substance mordanted in a natural extract of tannin. The process turns the patterned areas black. In contrast to the predominance of indigo dye techniques in West Africa and the southern Sahara, the complexity of the *bogolan* process stresses the importance of minerals in resist dyeing. Picton and Mack brought attention to this extraordinary process particular to Mali. It highlighted the relationship between a craft and its origins, signifying the environmental factors influencing its production – climate and local resources – and the sense of place that *bogolan* evokes in its cloth and material substance. The collective research of these scholars and museum curators has placed the foundation of modern textiles firmly in the context of Indian, South East Asian and African textile history.

In 1998 the British Museum Press published Jenny Balfour-Paul's expansive text on *Indigo*. In the 'Preface', Balfour-Paul described her role as 'a spider in the centre of a web, feeding from all sides' (1998: vii). She went on to describe the scope of disciplines embraced by indigo, from botany to chemistry, from ethnography to medicine, from economics to social history, accompanied by a breadth of references. In the twenty-first century, Dominique Cardon at the Centre for Scientific Research at the University of Lyons has developed extensive research into dyes obtained from plant, mineral and animal sources. *Natural Dyes: Sources, Tradition, Technology and Science* (2007) documents her intense scientific analysis of natural colour. Cardon has scoured museum archives, archaeological sources and historical documents to draw out the detailed multiplexity of dyes. Hereafter, this book develops new interpretations of the subject.

The emergence of a visual and sensory language

Following the survey of a literature that has been fundamentally descriptive and process orientated, this study goes on to examine the visual and sensory language that surrounds resist dyeing. From here, the review investigates difference in language – narrative, poetic, lyrical – in capturing the qualities of indigo. The research explores a range of personal voices in articulating the characteristics of resist dyeing, celebrating its many variables and conveying its energy.

O'Connor (2009/2014) recorded a 'Little Flowered Miao' skirt from Guizhou Province in South West China (Accession Number 2009.135.40) at the Pitt Rivers Museum, University of Oxford. Her commentary marks a distinctive shift in approach, a change that signals a new breadth and scope in description. O'Connor's handwritten notes reveal a personal system in the recording of historical and technical information:

> Wax resist. Top section is bought cotton from the market. Base is hand spliced hemp, woven, dyed etc. by hand. Pattern made by wax. Pattern v. complex and subtle. Indigo. Not purchased as one complete costume, bought from different people. If you are a married woman, you have different layers – not clear if this skirt is for a married or unmarried woman.
>
> O'Connor 2009/2014

In 2006 O'Connor donated her collection of 260 textile items from South West China to the Pitt Rivers Museum, Oxford. Assistant Curator Faye Belsey outlines their significance:

> During the 1980s and 90s O'Connor was collecting in a region that was only just becoming accessible to tourists. Most of the textiles had been bought mostly from villages, often directly from the makers or wearers, and at a time when textiles were not being produced specifically for the tourist market . . . this was an interesting period when the region was in touch with the wider world, while, at the same time, traditional methods and costume production were developing and changing.
>
> Belsey 2014

Consequently, in Belsey's view, '[i]t would be difficult, if not impossible, to go to South West China today and make collections of such range and quality' (Belsey 2014). Belsey stresses the importance of O'Connor's detailed notes accompanying the collection:

> The added richness of these assemblages comes from the expertise that O'Connor [was] . . . able to apply by recording detailed information about the individual pieces: the 'who, when, where, how, and why' that is so critical when objects are removed from their original contexts and placed in a museum.
>
> Belsey 2014

O'Connor's notes are observations and analysis from direct experience. They record a moment in time, an immediate response to an experiential encounter with people, places and objects. And while they reference context, social culture and provenance, they openly acknowledge unanswered questions. O'Connor's descriptions are empirical research as personal narrative.

In comparison, in his *A History of Dyed Textiles* Robinson describes knotting, folding, binding and ruching (known as *plangi* in India, *tritik* in Malaysia and

Indonesia, and *shibori* in Japan) as 'one of the oldest forms of pattern making on cloth':

> This would seem to be in part due to the basic simplicity of its techniques and equipment . . . also undoubtedly popular . . . because of the unusual results of the creeping dye and textural appearance left in the fabric after untying . . . The final result of any of these forms is a softness of blending of the design as the dye, despite the tightness of the ties, creeps along the yarn.
>
> Robinson 1969b: 78–9

Throughout his extensive historical and technical survey Robinson refers to the 'ingenuity' and 'inventiveness' in the application of resist techniques, and he stresses the influence of weave structure on their results, but acknowledgements of their particular visual and tactile qualities are scarce. Here, his use of terms such as 'softness', 'blending' and 'creeping' suggest the beginnings of an onomatopoeic or 'onomato*poetic*' representation of resist dyeing.

Importantly, the principles of resist have long been associated with natural dyeing, and the mutual empathy between the nuances and variations of resist and the subtle hues of natural colour are manifested in numerous examples of world textiles. In contrast to Robinson, when *Shibori: The Inventive Art of Japanese Shaped Resist Dyeing* (Wada, Kellogg Rice and Barton 1985) was reprinted in 1999, the website *thejapanpage.com* gave the following appraisal on the book's dust jacket: 'Potential for creating designs in textiles can be seen even in the physical properties of cloth. The simple fact that cloth tightly compressed into wrinkles or folds resists the penetration of dye is an opportunity – an opportunity to let the pliancy of textiles speak in making designs and patterns' (1999). Wada, Kellogg Rice and Barton describe the special characteristics of *shibori* as:

> [a] soft or blurry edged pattern . . . quite different from the sharp-edged resist obtained with stencil, paste and wax. With *shibori* the dyer works in concert with the materials, not in an effort to overcome their limitations but to allow them full expression. And an element of the unexpected is always present.
>
> Wada, Kellogg Rice and Barton 1985: 7

It is evident that this description begins to convey the *qualities* of resist dyeing, and the significant relationship between materials and processes and the actions performed on cloth. This book explores how the power of indigo – its distinctive alchemy and profound visual resonance – can be captured.

Frequently, the wonder of indigo and the fascination in its extraordinary qualities are expressed in florid descriptions. In her Introduction to *Indigo*,

Balfour-Paul describes her initial encounter with the dye: 'On a summer's day in 1981, in the courtyard of a Devon farmhouse, I first plunged my hands into an indigo dye vat, pulled out a length of cloth and watched the apparently magical transformation of yellow to indigo blue taking place before my eyes.' She describes being 'captivated' by the dye, referring to it as 'intriguing', 'extraordinary', with 'an aura of mystique', and Balfour-Paul's publication goes on to trace 'the myth and magic' of the dyestuff (1998: 1). In her chapter 'Blue Nails: Indigo Dyeing Worldwide' she describes the mysteriousness of a dye that only reveals its colour *after* the cloth emerges from the dye vat, declaring that even those with an understanding of chemistry 'cannot fail to be enchanted when they first see the slow transformation of yellow into blue'; the place where the indigo vat and oxygen 'turns sorcerer' (Balfour-Paul 1998: 115).

In 1996 the *Independent* published an obituary to Bosence. Crafts historian Margot Coatts wrote, 'For those who came under her powerful spell, colour was never quite the same again' (Coatts 1996). Balfour-Paul acknowledges Bosence as the dyer and printer who 'infected [her] with her enthusiasm' (1998: 131). It is most likely that her experience of indigo dyeing 'in the courtyard of a Devon farmhouse' took place with Bosence. In the Crafts Council's *Colour into Cloth* exhibition catalogue (1994), Barley Roscoe, founder curator of the Crafts Study Centre, quotes Bosence as saying, 'I feel my work has got something more than just making patterns. It's part of a vision or excitement, something you only catch on the wind of fleeting things, something you can't explain and you don't know where it comes from' (Roscoe 1994: 12).

This review of language in resist dyeing questions how the perpetual allure of the ancient substance of indigo can be articulated while striking a balance between expression and excess; poetry and sentimentalism. O'Connor's writings go further in capturing the sensitivity of expression in Bosence's work, presenting a real and more tangible interpretation. In 1982 in 'The Work is Part of Life' for *Crafts* magazine, O'Connor acknowledges that 'We do not have in English a neat phrase to describe the kind of artist Susan Bosence is' (1982: 42–8). Later, in her Foreword to Bosence's *Hand Block Printing and Resist Dyeing*, she describes Bosence's textiles as 'belonging to a world of natural beauty' (O'Connor 1985a: 8):

> They associate happily with stone, wood, plants. They look their best in natural light, creating a mood of quiet harmony. The cloths . . . have an apparent simplicity of pattern. Each mark . . . tells of the hand that has made it, so there is a vitality and subtlety at the same time . . . [Bosence aims to create] textiles that speak in their own terms and are true to their maker . . . One main message

> comes through . . . the use of dyeing techniques helps to develop an awareness of the cloth as a whole.
>
> O'Connor 1985a: 8

O'Connor viewed *Hand Block Printing and Resist Dyeing* as 'an invitation to share an approach'. According to O'Connor, Bosence had said that 'the final piece of cloth should be the whole piece of cloth – the quality of the cloth, the colour and the pattern of equal standard, lovely to look at, handle and use' (O'Connor 1985a: 8).

In parallel with metaphorical description, this book explores dyeing as performance, examining its actions and bodily engagement. Bosence's *Hand Block Printing and Resist Dyeing* represented 'hours of frustration trying to fathom the mysteries of colour and the art of dyeing' (1985: 12). Bosence's trials are echoed in historical accounts. In contrast to an idealized notion of indigo, they present a strikingly different view of the technical challenges and physical demands of its processes.

In 1813, in his *Philosophy of Permanent Colours*, Bancroft wrote a graphic description of a developing indigo dye vat taken from a report by botanist William Roxburgh. Bancroft felt 'induced to publish the work . . . which had been put into my hands at East India House' (1813: 174). Roxburgh's description illustrates the fermentation of indigo as a slow and animated process. He stresses that, as the plant substance slowly rots 'the peculiar *smell* of the plant has prevailed, but now it becomes very offensive, something like that of animal matter beginning to putrify [sic]' (Roxburgh cited in Bancroft 1813: 175). It is a description echoed in Robinson's reference to rotting fish in the Egyptian papyrus inscription (Robinson 1969b: 1). After considerable experiments, Roxburgh deduced that 'the most powerful precipitants added to these liquors cause no decomposition without the help of the open air'. Later, Roxburgh stressed that 'this *thing* was oxygen'. He was convinced that oxygen was 'the colouring principle of indigo' (Roxburgh cited in Bancroft 1813: 183).

In her study of *Indigo in the Arab World*, Balfour-Paul (1997: 87) notes that smell was not the only property of indigo to give offence. In Algeria, according to the traveller Touchon (1959: 118), 'even bones with slightly putrified [sic] meat attached to them were used to accelerate fermentation of the indigo vat'. Balfour-Paul notes that, in Tavernier's second volume of his *Travels to India* (1925: 9), the traveller records how the penetrative powers of indigo 'caused trouble to the processors in unexpected ways' (Balfour-Paul 1997: 87):

> Those who are employed to sift the indigo observe great precautions, for while so occupied they hold a cloth in front of their faces, and take care that all their orifices are well closed, only leaving two small holes in the cloth for their eyes . . . moreover, both those who sift the indigo and [those] who watch over them . . . have to drink milk every hour, this being a preservative against the subtlety of the indigo. I have . . . observed that if an egg is placed in the morning near one of the sifters [and] broken in the evening, it is found to be altogether blue inside, so penetrating is the dust of indigo.
>
> Tavernier 1925: 9, cited in Balfour-Paul 1997: 87

In Zabib, Yemen, during the 1980s, Balfour-Paul witnessed the universal and labour-intensive practice of treating undyed calico to bring 'cohesion and absorbency' to its substance in preparation for dyeing. Several 10-metre lengths were soaked in a cold solution of soda, then transferred to a large but shallow trough lined with plaster, before each length was moved to a smaller round trough and stamped vigorously, one at a time. Afterwards, each length of cloth was 'slapped hard against a stone column' and subsequently steamed for twenty-four hours by coiling each length around the rim of a pot full of water and soda, which was countersunk in a raised platform and heated from below. The cloths were then left to cool overnight before being stamped and beaten once more and finally rinsed (Balfour-Paul 1997: 84). Balfour-Paul's account resonates with the physicality of its processes, and her description exudes with the energy and force applied.

In *The Dyer's Handbook: Memoirs of an 18th Century Master Colourist*, Dominique Cardon (2016) translates a French recipe of the time for achieving black on cloth. Its anonymous author documents the extraordinarily protracted process. It consists of concentrated tannin, extracted from gallnuts and logwood, dissolved in a bath of madder dye. Blue cloth previously dyed in woad is immersed in the solution of tannin, followed by subsequent dyeings in more madder, ferrous sulphate ('copperas') and weld. Here, the Frenchman's recipe reflects the intensity of his performative procedures – the heavy weight of the cloth, the heat of the dye vat, the constant stirring, the repetitive lifting – the physicality of multiple processes spanning day and night:

> *Woaded Blacks*
>
> Let the cloth be given a woad ground of an azure blue . . . [Place] powdered gallnut and logwood in a madder bath . . . Simmer for two hours. Pass the cloths in this galling bath [tannin] during one and a half hours, boiling constantly . . . Heave out and air well . . . Then dye common madder, and into the same bath . . .

> put copperas [iron sulphate] previously dissolved before being poured into the vessel. Turn the cloth in it, heave it out, and air well. Give a second dip. Lastly, heave out the cloth, and put it back into the vessel, where it must spend the night. Observing that there should not be any fire left burning in the furnace, even remove the embers before saddening. The next day, heave out the cloth, start having it scoured of the copperas in the stock for half an hour. Let it be washed another time with soap if the fuller is good at scouring it out afterwards. After this operation, pass it in a strong weld bath.
>
> Cardon 2016: 100–1

William Morris is acknowledged as the principal proponent of natural dyes in the Arts and Crafts movement of the nineteenth century. In her study of *The Wardle Family and its Circle* (2019), Brenda King describes the significance of Morris's relationship with Thomas Wardle – the authority in natural dyeing at that time: 'Theirs was an alliance of like-minded polymaths and, as with many collaborations, it depended on compromise and the willingness to make mistakes for the sake of progress' (King 2019: 8). King notes that Morris made several visits to Wardle in Leek, Staffordshire 'to learn the latest advances', and she describes the two men as 'well matched in their restless search for almost forgotten techniques' (King 2019: 9).

In Gillian Naylor's compilation of Morris's writings, *William Morris: by Himself* (1988), on 28 March 1876 Morris writes to his friend Aglaia Coronio about indigo dyeing with Wardle, 'working in sabots and blouse pretty much all day long' (Morris 1876a, cited in Naylor 1988: 130). Morris's letters reveal a picture of frustration and his endurance of the physical demands of indigo dyeing. Morris writes to his confidante Georgiana Burne-Jones on 4 February 1877:

> I have been dyeing in *her* [the blue-vat] all the afternoon, and my hands are a woeful spectacle in consequence . . . I lost my temper in the dyehouse for the first time this afternoon: they had been very trying: but I wish I hadn't been such a fool; perhaps they will turn me out tomorrow morning, or put me in the blue vat.
>
> Morris 1877a, cited in Naylor 1988: 133

Three days later, he writes to his wife, Janey Morris:

> Please I shall want a bath when I come home: you may imagine that I shall not be very presentable as to colour . . . We had to work it at 130° [Fahrenheit] and a hot work it was as you must keep the goods clean under the surface . . . I set myself too much work to do: that is a fact.
>
> Morris 1877b, cited in Naylor 1988: 133

Following the introduction of synthetic dyes, Morris's collaboration with Wardle led to the rediscovery of indigo in the Arts and Crafts movement, but, in a letter to Wardle on 9 May 1876, Morris describes the process as 'a troublesome affair'. Proposing to transfer his production from Leek to Hammersmith, Morris declares, 'I have no doubt I should flounder about finely in the blunder-sea.' Morris writes with a sense of commitment coupled with impending despair (Morris 1876b, cited in Naylor 1988: 131):

> As to going on with the enterprise . . . I am determined to do so in some way or other; because so much of my work depends on the solving of the difficulty, that no amount of money could compensate me for the disappointment of having to give it up: at the same time I don't suppose the dyeing of our wools will ever be a profitable business to anyone, and no doubt it will be a troublesome one.
>
> Morris 1876b, cited in Naylor 1988: 131

On 30 November 1877 Morris writes to Wardle:

> The difficulties of your carrying on a private manufactory for us, a business which must in the nature of things be rather what is called today an art rather than a manufacture [is] I feel, to be immense.
>
> Morris 1877c, cited in Naylor 1988: 133

Just as Morris and Wardle restored much of the knowledge of natural dyes in the nineteenth century, Barron and Larcher rediscovered them in the twentieth century. In the Crafts Council slide pack of Barron and Larcher's work (Bosence and Roscoe 1978), according to Roscoe, it was Barron's 'avid' experimentation with processes that informed her understanding (1978: 3). Bosence recalls Barron finding out 'often by dangerous trial and error' (1978: 7). Barron returned to the principles of eighteenth- and nineteenth-century books on dye chemistry in the libraries of the British Museum, the Victoria and Albert Museum and the Patent Office (Roscoe 1978: 3), and her study of them continued to inform her experiments.

In 'My Life as a Block Printer' (1964), the transcript of a Ministry of Education talk at Dartington College now in the Crafts Study Centre, Barron said:

> There's something magic about indigo. It is a fermentation, like wine, beer and lots of other good things . . . It ferments more easily in a hot country – in India it stays gently bubbling with 'white froth' . . . Here it has to be kept tepid artificially. So, when I had made my vat, as the weather was rather cold, I wrapped the barrel in some old blankets and put it near the old anthracite stove . . . I couldn't stay

> and watch it all the time between the intervals of stirring, so one night I left it and went to a party.
>
> Barron 1964: 50

When Barron returned, the stove had set fire to the blankets, burnt a hole in the base of the vat, and the fire had been put out by the lost indigo: 'I had dyed practically nothing in it, and there was a great mess to clear up.' Having read that urine was better than lime in the recipe, Barron collected buckets of the stuff from her neighbours at night, resulting in a 'disgusting smell' in her Hampstead street (Barron 1964: 50–1).

On their webpage *Dyeing Now: Contemporary Makers Celebrate Ethel Mairet's Legacy*, Ditchling Museum regard Ethel Mairet's *A Book of Vegetable Dyes* (1916) as 'the definitive text on natural dyes'. Mairet was a pioneer of twentieth-century craft, and Barron described working with this 'marvellous weaver and spinner'. Barron wrote of Mairet's 'dye weeks' where she 'got various people to come and dye stuff for her to weave'. Barron also described Mairet as 'not a terribly good dyer' but someone who was 'clever enough to get people to come and dye what she needed [for weaving]' (Barron 1964/2018: 53).

Poetic language

This book investigates how the physical making of resist-dyed textiles, its dynamic forces and slow expression can be interpreted as performance and bring new meaning to contemporary cloth.

Jane Weir explores the performance of craft through experimental writing. In *Walking the Block* (2008) she reflects on Barron's practices and presents a narrative of the work through poetry; observing colour on cloth, capturing technical notations, and reinterpreting them in her own voice. On their webpage *Jane Weir*, the Poetry Archive classifies Weir's work as 'poetic biography', describing it as 'distinguished by its sensuous handling of surfaces and textures; the simultaneous description and practical making – through language – of patterns and forms'. Weir's poem 'Preparation In The Kitchen, Attempt At An Indigo Vat' (2008) reflects upon Barron's early explorations: 'Now all to do was spread the word amongst my friends, put in the post a bottle and a note saying – *I need night urine, in buckets provided please! Ignite the flame, stand over – don't stop stirring*' (Weir 2008: 28).

Bosence stresses Barron and Larcher as 'first and foremost artists' with 'an intense commitment to, and love of, their work . . . This helped them to overcome tremendous technical and often physical problems' (Bosence 1978), as expressed later in the same poem (Weir 2008: 28):

> Waited.
> Waited, for the dark blue iridescent scum, its clear yellow liquid coiling below.
> Avoid air in vat. Pour without splashing.
> Stir.
> In another pail I housed a solution for wetting, composed of some sort of neutral detergent such as teepol, to one gallon of warm water.
> Wetted.
>
> Weir 2008: 28

In *Jane Weir*, the Poetry Archive describes Weir's writing as 'meticulous, carefully inquiring and quizzical; clear and even, so as not to linger on a particular phrase or constituent part, but rather to keep a sense of the "weave and pattern", the wider tapestry'. Here, dyeing is reimagined in Weir's 'When The Indigo Flowers' (2008: 34):

> High summer.
> I stare into the indigo vat,
> sloshing,
> a cabochon of deep
> delphinium blue,
> in mint condition
> like a brand new dress
> before the tag's torn off.
> Nothing can prepare you
> for the accumulation
> of snorting bronze bubbles
> that pitter patter across its dense centre, or the tiniest light blue bubbles
> that complete
> the shape of two tiaras crowning; nothing can compete.
> I serve the flower – holy with two hands.
> In awe I skim the flower
> before baptising blue, a length of wetted cloth.
>
> Weir 2008: 34

Weir's literary examples illustrate the physical demands of dyeing while expressing its dynamics, applying poetic language to translate the intensity of its processes and the creative energy employed in its practices.

The space and place of resist dyeing

O'Connor and Weir step back from the process and notate their observations, using their personal voice to convey the animation of dyeing and the power of its art. Together, the work of the writers, historians, ethnographers and practitioners discussed also evoke a sense of place in their accounts of different cultures and applications, both historically and in modern and contemporary practice. The perceptions of space and place that unfold in their narratives locate resist dyeing in time and real experience.

The anonymous author of *The Dyer's Handbook: Memoirs of an 18th Century Master Colourist* does not divulge the location of his manufacture. But he gives details of the practical conditions in which he conducts his experiments, commenting on the quality of water resources vital for dyeing. In her translation of the author's handbook, Cardon's research pays considerable attention to the rivers of the Languedoc and the importance of its regional water supply. The dyer goes on to reveal that the place where his colours were produced had two different water supplies for different purposes: 'For dyeing and washing the cloths, a river that receives additional water from a group of springs a short distance upstream is used, and another spring is directed to a fulling mill' (Author cited by Cardon 2016: 15). Cardon explains:

> A spring called La Bouillette still exists although it more and more frequently dries up, like many other springs in Languedoc. In that same place, a river, the Cesse, flows right at the foot of the high walls of a former Royal Manufacture, and a few miles upstream, a group of springs below the Palaeolithic caves of La Fons, mingle their tepid waters in a whirl of bubbles at the bottom of a mossy rock basin, before overflowing into the river.
>
> Cardon 2016: 15

Cardon's discovery was a 'breakthrough'. It enabled her to identify the most likely location of the author's manufacture. The description of his working environment proved to be key in determining the place particular to his practice.

Similarly, Balfour-Paul's knowledge is the result of direct experience. Extensive field study is the cornerstone of her research. Through systematic investigation among world communities and in international collections, and through consultation with specialist academics and practitioners, Balfour-Paul has built an expansive collection of textual records, enhanced by the richness of her photographic observations. Balfour-Paul's resolution to research indigo dyeing was 'reinforced' by her personal experience of it in Yemen and parts of the Near and Middle East, where practices were 'largely unrecorded' and 'about to disappear' (Balfour-Paul 1998: vii). Balfour-Paul's research captures the richness of its history and the development of its production through global trade. Her comprehensive international survey of indigo outlines some of the spiritual associations with its substance and its many cultural connections.

It is in her chapter 'Blue Nails: Indigo Dyeing Worldwide' (1998: 115–45) that Balfour-Paul pinpoints unique processes particular to specific parts of the world; practices that not only confirm the significance of indigo, but demonstrate its importance in the definition of place. According to Balfour-Paul, the Amazonian Indians spat masticating indigo leaves and lime directly onto bark cloth, working it in with their fingers, and they discovered the value of saliva by chewing a local fruit and extracting a kind of indigo from its juices. She notes that the easiest and simplest method in the fermentation of natural indigo is illustrated in the vats of composted material in Japan where the addition of local rice bran, rice wine, honey or lime speeds up the fermentation process through to maturity. This contrasts with the process of drying and forming balls of indigo in Nigeria, or compacting it into blocks of 'cake' as practised in other parts of the world including Europe, where the dyestuff is imported, stored and later broken down to be wetted again and regenerated in a new dye vat.

In contrast, Bosence's world in textiles was intimate, contained within the boundaries of home surrounded by a familiar network of people. At her memorial in 1996, her son, Daniel Bosence, described the block printing and dyeing carried out in the house as 'absolutely normal'. He said: 'As a child, I had known nothing else. Indeed, when I visited my friends' houses these appeared strange. I could see the dining room table but I wondered where they kept their block printing table' (Bosence D. 1996: 23–4). Bosence developed her work in her domestic space and through a close attachment with mentors, colleagues and students. Her textiles reflect her acute sensitivity to materials and processes, her understanding and awareness seemingly intensified by her domestic environment. While Bosence's practice might appear sheltered, protected within the certainty and security of home, her working environment was her bedrock. Her personal identity and strong sense of place are characterized in the simplicity of her work. In her concluding chapter on 'Dyes' in *Hand Block Printing and Resist Dyeing*, Bosence wrote:

> I have been strongly influenced by those colours occurring naturally around me in this isolated part of the country, so I feel diffident about using strong, vibrant colour schemes . . . I do feel this [in Southern Europe] when I am there. Similarly

> . . . a visit to London by train or car amid the crowds, buildings, traffic lights, flashing signs . . . and all that goes with metropolitan life. But I have lived in Devon long enough to feel that mine will not change.
>
> Bosence 1985: 24

Bosence's knowledge of resist-dyed cloth covered a broad field of world textiles. Her sensitivity to substance was the result of an accumulation of personal encounters with materials and processes in her own space and within the familiarity of her environment; a continuous experiential flow that defined her place.

It can be seen, therefore, that writings on indigo are rich with textual embellishment that recur across different genres. Furthermore, the practice of narrative to communicate craft has emerged in the subject of textile dyeing.

Narrative language: emotional connections

In 2010, artist and author Edmund de Waal wrote his memoir of a collection of 264 Japanese *netsuke* in *The Hare with Amber Eyes: A Hidden Inheritance*. The *netsuke* – tiny carvings worn as ornamental toggles in the traditional dress of Japan – formed 'a very big collection of very small objects' (de Waal 2010: 10). He writes:

> Most of them are signed . . . When you tumble them in your hands there is a pleasure in finding where these signatures have been placed . . . as well as the play between the strokes . . . I intermittently roll this thing through my fingers. I realise how much I care about how this hard-to-soft, losable object has survived. I need to find a way of unravelling its story.
>
> de Waal 2010: 10–13

De Waal retraces the journey of his collection across nations and continents, following the pathway of five generations of his family through the drama and tragedy of world events. An emotional connection with the materiality of these objects – wood, ivory, amber – reconnects de Waal with portrayals of the past. Each *netsuke* held in the hand embodies memory, each form comes to represent an evocative vignette of social history. De Waal's storytelling through objects marks a turning point in craft literature. His use of narrative as a means of communicating the sensory power of material substance was groundbreaking.

Since then, Cardon and Balfour-Paul have published historical memoirs on dyeing, each introducing elements of expressive language within their informative text. Cardon's research into natural dyes, including woad and indigo, is based on the records of past masters and the notes and narratives of their craft left behind. In 2016, in *The Dyer's Handbook: Memoirs of an 18th Century Master Colourist*, Cardon developed an understanding of the daily preoccupation of a textile dyer of broadcloth. By stepping into the author's shoes, she reinterprets the documents of his diaries in the Languedoc region of France during the eighteenth century. In her 'First Encounter' she describes how the manuscript was revealed to her from a desk drawer: '. . . a high, thin notebook with a frail, partly torn, cream-coloured paper cover. I opened it – or rather, it opened itself – at the middle page: on both left margins, a column of colour names; just beside, petals of thin, velvety cloth of different . . . shapes and shades.' Facing the swatches, on the right column of either page, were the author's explanations on how he had made each hue. For every special shade of green, the exact degrees of woad and indigo blue ground were clearly defined (Cardon 2016: 13).

Cardon's engagement with the original memoir is personal. She refers to it as a 'maimed manuscript', and describes the tactility of its material content. Cardon observes 'the mode of pagination' in its 100 pages: 'The layout is clearly devised to make it look like a printed book.' This was typical of the documentation of arts, science and literature in the eighteenth century, and in keeping with the spirit of the Age of Enlightenment (Cardon 2016: 33). Cardon records the 'secrets' held in the traces of its history, how the memories of its lost sheets are revealed in the marks left behind from the cut of a quill knife, and how the only remaining means of identification is in the 'List of contents'. The manuscript carries clues to the restoration of its dislocated pages, the remaining ones bearing stitches and earlier repairs made with strips of paper and glue. There are dyed cloth samples that have been cut, while 'other samples have been entirely peeled off the page, only leaving a few coloured fibres stuck in remnants of glue, or even tearing a hole in the page'. And there are mistakes in the text 'characteristic of someone who either is writing under dictation and has not heard or understood well what was dictated, or is trying to copy draft pages hastily written in a handwriting hard to decipher' (Cardon 2016: 33–7).

Cardon's account tells the story of the manuscript, its life, and her immersive passion with its pages. At the close of the section 'First Encounter', Cardon declares: 'All the pages . . . are bloomed with delightful assortments of vivid or subtle colours and detailed recipes for their creation.' On her first encounter with the manuscript, she wrote, 'That summer afternoon, I fell in love with the document.' Cardon resolved to publish and share it 'with other lovers of colour' (2016: 14). In the Foreword she presents us with an invitation:

> Imagine. Colours of the past, escaping from the pages of old dye books and pattern books. Persian blue, raven, dainty blue, pomegranate flower, spiny lobster, winesoup, pale flesh, dove breast, golden wax, grass green, green sand, rotten olive, modest plum, agate, rich French gray, gunpowder of the English, finding their way to the streets of our cities, enlivening everything we wear, all allied to dissipate the bleakness of the times. 'You may say I'm a dreamer' – historians do like to think that understanding the past can inspire the future.
>
> Cardon 2016: 11

Similarly, Balfour-Paul has drawn parallels between her own research into indigo and that of earlier explorers. In 2015 in *Deeper than Indigo*, Balfour-Paul entwined her own experiences of indigo in different regions of the world with Thomas Machell's travelogue of indigo production recorded at the height of British colonialism in the nineteenth century. In an interview with Stuart Jeffries (2015) in the *Guardian*, Jeffries suggests that Balfour-Paul's initial idea was 'to edit the journals of the man she had so fallen for, and publish them with an introduction and annotations' (Jeffries 2015). Machell worked in Bengal, where indigo dye was made by adding oxygen to vast vats full of green *Indigofera* leaves. The resultant liquid was dried into hard blocks, a process that Machell oversaw. Balfour-Paul deduced that Machell was afflicted with some kind of infirmity, most probably lameness. As a result of his disability, Machell did not fit easily into Victorian society, but, in Balfour-Paul's view, 'It helped that he was an outsider. It's what makes his journals so richly observed' (Balfour-Paul cited in Jeffries 2015). In contrast to Cardon's approach, Balfour-Paul interweaves her own travels with those of Machell, perceiving herself and Machell to be two kindred spirits, and writing about the places they had both visited. In writing her journal she confides in Machell, and was even considered to be his reincarnation by the Indian psychologist Rashna Imhasly-Gandhy (Balfour-Paul 2015: 205). In Jeffries' view, 'these transcendental passages serve the book's narrative well, since they can be read as [Balfour-Paul's] imaginative projections of what Machell may have written in the last two volumes of his journals – volumes that have never been found, if indeed they were ever written' (Jeffries 2015).

Both Cardon and Balfour-Paul demonstrate a deep emotional connection with their respective subjects. While connotations with love affairs and romanticism have been suggested, more importantly, Cardon and Balfour-Paul's works illustrate their shared empathy with writers of a different generation. They express the enduring passion for colour on cloth that has been sustained throughout history since inscriptions on dyeing were first found on the ancient papyrus referred to by Robinson (1969b: 1). Significantly, these works use evocative language to express a profound relationship with dyers of the past.

The language of intangible cultural heritage

In 2002, Wada introduced the term 'memory' in her publication *Memory on Cloth: Shibori Now*. Wada uses 'memory' to express both the patterning and the three-dimensional forms resulting from cloth manipulation: 'Cloth sensitively records both the shape and the pressure.' Wada perceives 'memory' in the shapes and forms imprinted in the cloth: 'This is the essence of *shibori*' (Wada 2002: 8). 'Memory' in *shibori* equates to process and what it leaves behind; what is left after the manipulations have been worked and unravelled. In parallel, resist dyeing has its own culture, a place where memories are embedded over time. In the context of resist dyeing, 'memory' is both physical and metaphorical, and the notion is entwined with intangible cultural heritage in the subject.

In 2003 the United Nations Educational, Scientific and Cultural Organization (UNESCO) established a *Convention for the Safeguarding of Intangible Cultural Heritage*. Article 2 of its text proposed five broad 'domains' for classification spanning oral traditions and expressions; performing arts; social practices and rituals; knowledge and practices in nature; and the fifth element, 'traditional craft' (2003a). UNESCO's objectives for safeguarding any traditional craft include two primary areas of focus – the transmission of knowledge and the continuity of its practices (2003b). In defining 'intangible cultural heritage' UNESCO states:

> While fragile, intangible cultural heritage is an important factor in maintaining cultural diversity in the face of growing globalization. An understanding of the intangible cultural heritage of different communities helps with intercultural dialogue, and encourages mutual respect for other ways of life . . . The importance of intangible cultural heritage is not the cultural manifestation itself but rather the wealth of knowledge and skills that is transmitted through it.
>
> UNESCO 2003c

UNESCO's *Text of the Convention for the Safeguarding of the Intangible Cultural Heritage* (2003a) includes recurring themes, such as 'education, training, and development', 'raising awareness and promotion' and 'fostering scientific, technical and artistic studies', as a common vocabulary. Contemporary engagement is referred to as 'international cooperation', 'the exchange of information and experience' and 'joint initiatives'. While the language is inclusive, there is a notable absence of imaginative vision reflected in these phrases. Generalized terms are painted with a broad brush.

UNESCO acknowledges that 'Traditional craftsmanship is perhaps the most tangible manifestation of intangible cultural heritage' (2003b). The traditional practices of resist dyeing are recognized largely through their products. Textile terms are defined by their visual characteristics – the diffused shapes and linear grounds associated with Japanese *shibori*; the floral and geometric repeats of white on blue associated with European 'blueprint'; the broken wax and crackled textures associated with Indonesian batik. Literature on resist dyeing is strewn with descriptions of the tangible – what is seen and can be documented. Expressions that draw out the intangible qualities of resist dyeing are less evident. Identification by their end product conceals other qualities and circumstances unique to them, restricting the potential to reveal new interpretations and broader awareness and understanding in the longer term. 'Blueprint' holds memories in the long history of resist printing and indigo dyeing in Central Europe. In the words of Vydra:

> A single colour – indigo – together with a hand-printing process has already for a thousand years provided so much artistic beauty and variety, such a fruitful field of artistic creation, that is certain to survive, in various forms, for many a future generation.
>
> Vydra 1954: 7

By 1985 only one artisan continued to produce 'blueprint' in Slovakia in accordance with traditional practices. Vydra's *Indigo Blue Print in Slovak Folk Art* was published in 1954 during the Soviet occupation of Czechoslovakia. In post-Second World War Slovakia, 'blueprint' was supported by the state-organized ÚĽUV, the Centre for Folk Art Production, in Bratislava. It is notable that Vydra refers to the Slovak dyers as 'devoting their work solely to the service of the people and its needs' (1954: 24). Perhaps, in some part, this explains why the intangible cultural heritage of European 'blueprint' is defined in the tangibility of its processes and products.

While Vydra describes 'blueprint' as 'artistic', its poetic expression rarely occurs within its literature. In her survey *Blueprint in Slovakia* (2014), Oľga Danglová identifies a 'unification of the language of decorative blueprint design' (2014: 357). She attributes this to the obligatory journeys that dyers were required to undertake during the seventeenth century. After the completion of a three-year apprenticeship, dyers had to travel for a further three to nine years, working in different communities across Central Europe to master their practices before they could be fully accepted into their local guilds. Vydra refers to the Slovak people as 'devising their own nomenclature' – a system of names for designs that included 'pear', 'vine', 'cleft walnut', 'lace', 'sun', 'tow', 'pincer', 'thumb' and others (1954: 32). Danglová attributes this 'language' to 'the rich reservoir of patterns' that were acquired and exchanged at that time, and how they have been 'mixed, compiled and recycled according to contemporary demands'. According to Danglová, these names demonstrate the 'common roots manifested in the 'compositions and ornaments' of 'blueprint', and the interrelationship between different communities (2014: 357).

In contrast to the descriptive language of resist dyeing in Central Europe, expression in Japanese culture is significantly poetic, and this is exemplified in the language of colour. In Shigeki Fukumoto's *Urges/Urgings: Notes on a Theory of the Art of Japanese Dyeing* (2016), the author explores the central importance of the Japanese sensibility of *shimi* (deep penetration) and *shimijimi* (keenly felt emotion), and their significance in dyeing that is 'so often at the core of Japanese art' (2016: 7). Fukumoto explains that, in Japanese, meanings of 'to feel in the heart' and 'to dye' can both be expressed by the verb *shimiru* – to soak in or permeate. Both terms express the permeation of water or dye, and the sensation of inner feelings (2016: 31). In the recording of *In Conversation: Lesley Millar and Reiko Sudo* (2020), textile designer Reiko Sudo discusses her association with a Japanese heritage indigo dyer in the making of *Ogi No Mai / Japanese Fanfare* (Sudo 2020), her installation in the exhibition *Fabric: Touch and Identity* at Compton Verney, Warwickshire. According to Sudo, the indigo dyer worked with water sourced from the natural spring on his estate. She described the long process of oxidation in which the dyed cloth was left to soak under a slow-flowing tap for one month, a slow permeation of water and a continuing expression of *shimiru* exemplified in contemporary indigo dye practices.

Architect, designer and writer Kenya Hara introduces his publication *White* with the words, 'This is not a book about colour' (2007: i–ii). Rather, he describes it as an attempt 'to find the source of a Japanese aesthetic that produces simplicity and subtlety through the concept of white'. Hara asks, 'What is colour?' In his view, while classifications of colour explain chroma and hue, they do not represent 'a full sense of what we perceive as colour'. Hara maintains that colour is not understood through our visual sense alone, but through all our senses: 'Insofar as colour systems are based solely on the physical [and] visual nature of objects, they cannot convey our total response' (Hara 2007: i–ii). Significantly,

Hara explains that in eighth-century Japanese poetry, only four basic adjectives were used in relation to colour. They were derived from *akai* (red), *kuroi* (black), *shiroi* (white) and *aoi* (blue). With the addition of 'i' to these nouns, they became adjectives expressing a state of brightness and energy in red, the absence of light in black, brilliance in white and an impression of obscurity in blue (Hara 2007: 5):

> These four adjectives may strike us as too few, but . . . each word had a very broad range, and people were able to express subtle differences in meaning and atmosphere . . . There was no need to classify colours as strictly as we do today . . . blue and green could be lumped together emotionally, under the broad category of blue (*aoi*) . . . communicating colour through . . . the names of plant-based dyes such as indigo.
>
> Hara 2007: 5–6

Hara refers to a sample book titled *Traditional Japanese Colours*, organized according to traditional names with 'the power to stir one's imagination' (Hara 2007: 3):

> When I pick up this book, I can accept the nature of the colour captured by the words smoothly and naturally. At the same time, my senses are awakened; I feel soothed as if hearing the dialect of my hometown, as well [as] a tinge of loneliness.
>
> Hara 2007: 4

Hara's observations correspond with memory, and the intangible cultural heritage embedded within colour. Words and names transport him to experiences of the past. Hara points out that 'there are probably as many systems of colour in this world as there are languages or cultures. "Traditional Japanese colours" is just one of them' (Hara 2007: 7). While Central Europe has devised systems derived from a visual understanding of resist dyeing, Japanese culture has developed an inherent understanding through other senses. In their different ways, Europe and Japan have developed terms that begin to explain the varying interpretations of intangible cultural heritage – how it is identified and defined.

In contrast to European names for patterns based on the visual depiction of objects and natural forms, Hara's historical references define the relationship between colour and the senses. Expressions of indigo, in all its hues from blues to greens, are implicit within the intangible cultural heritage of resist-dyed textiles. Hara describes this understanding in metaphorical terms, associating it with an accumulative process over time: 'Just as a stalactite cave is formed through the accumulation of droplets that fall one by one in dizzying repetition over time, so mental images of the brilliance of nature or the changing world gradually accumulate to form the names of colours' (Hara 2007: 7).

In conclusion, Hara observes an understanding of colour embedded in intangible cultural heritage: 'Some things are lost and others transformed, but finally, without anyone being aware of it, colour becomes established as a grand system of consciousness' (Hara 2007: 7).

Unlike European words that describe an object of nature, Japanese words capture a moment in nature. In Hara's view, words such as *moegi-iro*, 'the bright green of budding plants', or *asagi-iro*, 'the greenish blue of the leaves of the leek', are fragile but have the power to convince because 'they capture the moment of observation. This is why they are so deeply felt' (Hara 2007: 6). Hara's description of emotion is a metaphorical expression of intangible cultural heritage through the elements of thread and cloth:

> The names of colours function like a thread attached to a frightfully slender needle, capable of stitching together our most delicate emotions. When the needle hits its target, we feel either pleasure or empathy. It can also strike us as very painful, since it makes us aware that this delicacy is disappearing from our modern living environment.
>
> Hara 2007: 6

The breadth of approach in the expression of resist dyeing paves the way for new interpretations. This book builds on the existing language surveyed in this review, but takes it in a new direction. It examines the theories of key thinkers in other disciplines who use metaphor and notions of performance to develop critical discourses within their subject – anthropology, cultural theory, geography. The inquiry considers resist dyeing from different perspectives. It identifies images and concepts that resonate in the materials and practices of resist dyeing, and evolves towards an exploration of ideas in fresh contexts – art and architecture, movement and motion, space and place. In her chapter 'A final word' in *Indigo* (1998), Balfour-Paul concludes: 'Certain words contain within them, like Russian dolls or pass the parcel packages, layer upon layer of meanings as the wrapping is peeled away' (1998: 233). From here, this book seeks to develop a new understanding of resist dyeing through metaphorical and performance meanings in modern and contemporary cloth.

2

Movement, Motion and Time: The Formation of Practice

In 2009, speaking in Amsterdam at *Showing Making: International Conference on the Representation of Image Making and Creative Practices in Ritual, Art, Media and Science*, anthropologist Tim Ingold defined the mechanics of movement and the formation of knowledge as a place where 'cognition builds up into an integrated assembly [of points and materials]'. He describes this phenomenon as 'a path traced through the terrain of lived experience; a meshwork of interwoven trails' (2006: 30).

This chapter explores the practices of artists and makers who have contributed to indigo and resist dyeing during the twentieth century. It identifies the routes and pathways critical to their work, considering the influence of the domestic workshop and professional studio on their making. With reference to the geographer Yi-Fu Tuan's concepts of space and place, the chapter questions how working spaces and systems of operation, past and present, signal the importance of pattern, rhythm and repeat in resist processes, and how they have come to influence the development of resist dyeing in modern and contemporary cloth. The philosophies and practices of English dyers and printers are compared with their counterparts in Japan, and the chapter goes on to examine how cultural connections are revealed in cloth, both in the formation of the practices that produce them and in the orientation of the workspaces where they are made.

Personal accounts of field studies in Central Europe and Japan consider the movement and motion of practice in real time. A continuous 5-metre length of hand-drawn, wax resist dyed linen is introduced as an expression of flow, and the chapter concludes with a narrative on the making of *Rhythm* (Brassington 2017), a piece that became a configuration of movement, motion and time on cloth.

The domestic space: routes and pathways to making

According to Simon Olding, former Director of the Crafts Study Centre, whether the studio 'looks like home, a classroom or a factory' it is a place where tools, ideas and characters are set for transformation (2012: 10). Susan Bosence is a principal figure in twentieth-century resist-dyed textiles. In *Hand Block Printing and Resist Dyeing* (1985), Bosence wrote, 'I had become seduced by indigo resist dyeing; tying up tiny spots, away from the workshop, out in the sun or beside the fire' (Bosence 1985: 77). Her workshop was in a converted barn attached to her farmhouse in Sigford, Devon and, in Deryn O'Connor's view, 'the scale of Susan's work is domestic. . . life and work are inseparable. Hens, geese, dogs, grandchildren [are] kept outside by the lower half of the stable door. The bread may as well be rising on the workshop Rayburn as on the Aga in the kitchen' (O'Connor 1982: 48). Describing her workshop, Bosence wrote that the space was 'quite large enough to heat during winter'. The stove, 'be it Aga, Rayburn or woodburner', seemed 'essential to heat the room and the water, cook the dyes, air damp cloths, and boil a kettle for tea'. Rainwater was stored in the garden and in the workshop, and in the warmer months, when 'life was quite different', the work was done outside with dye vats, hanging lines and tables (Bosence 1985: 74–7). In the view of Yi-Fu Tuan, 'The home itself feels more intimate in winter than in summer. Winter reminds us of our vulnerability and defines the home as shelter. Summer, in contrast, turns the whole world into Eden, so that no corner is more protective than another' (Tuan 1977: 137).

O'Connor observed that 'Susan's work does not lend itself easily to exhibition . . . [While] some craftwork today seems more at home in a gallery with bright lights than anywhere else; Susan's textiles are the reverse . . . they ask for natural light and air' (O'Connor 1982: 48). In her Foreword to Bosence's publication, O'Connor wrote, 'She herself has not sought the limelight . . . Susan's work is in no way "museum art". Her textiles are made for use, for clothes or in the home. Their scale is chiefly domestic and they belong to a world of natural beauty' (O'Connor 1985a: 8).

Bosence was mentored by Phyllis Barron and Dorothy Larcher, who are recognized as two of the leading textile dyers and printers of the 1920s and 30s in England. Barron and Larcher's activities and approaches to making span both domestic life and professional practice. Barron only began experimenting with dyes when she moved into her first-floor flat in Hampstead, London. She started by amassing bits of equipment, tubs and basins (Bosence 1985: 42). Later, when Bosence was taught by Barron in the processes of indigo dyeing, Bosence recalled 'much to-ing and fro-ing' to find measures, jugs, sacks and hot water: 'Ever since that day, there have been years of excitement, frustration and fun from sharing the experience' (1985: 116). Bosence recounted indigo dyeing as a communal event, a domestic activity to be 'shared with friends' (1985: 118). She wrote, 'We all knew more or less what to expect – that we shall finish with blue hands, faces and feet, but there will be indigo cloths for all' (1985: 116). Jane Callender was a student of Bosence. She describes how 'living in a maisonette in Beckenham, I sat the small indigo vat at the end of the bath, and fabrics dripped into buckets on the balcony or on friends' gardens' (Callender 2017: 176). Now, at her home in Norfolk, Callender's indigo vats have moved to the garden, and in the warmer months she considers 'the outside to be my studio' (Brassington 2015: 64). The greenhouse becomes her dyehouse (Callender 2017: 153).

O'Connor explained that, as well as indigo, Bosence used a modern form of vat dyes (Soledons) that required her block printed cloth to be passed through a bath of sulphuric acid solution to fix the printed colours. This was followed by repeated rinses. Similarly, her mineral dyes needed to be 'raised' in a caustic soda bath to develop their colour, while another vat dye (Caledon) and indigo were built up through 'a series of immersions and airings in succession' (O'Connor 1982: 44). Reflecting on the work of Bosence, O'Connor wrote:

> All the methods Susan uses have one thing in common – they all involve immersion of cloth in liquid. Nowadays, with launderettes and washing machines, some people may not know how much cloth weighs when wet, nor have experienced the repeated raising and lowering necessary to allow liquid to penetrate all parts of a cloth. It is an eye opener to watch Susan handling cloth in water, as she firmly coaxes it to receive dye or chemical.
>
> O'Connor 1982: 43–4

Tuan observes that physical space is measurable by the reach of one's arms. In his view, the sensation of outstretched arms 'enlarges a person's world'. Space expands, reaching out to new physical and creative horizons (Tuan 1977: 53). According to O'Connor, Bosence did not design 'graphically': 'What happens between her hands and the piece of cloth is the action of designing' (1982: 44).

Bosence described the motion of indigo dyeing at her workshop, a place integrated within the environment of home. Baskets were filled with prepared cloth. She wrote, 'Take a length, find a selvedge and, starting from one end, run your thumb and finger along it. Lower the cloth . . . imagine swimming under water . . . think of the movement of your hands' (Bosence 1985: 119). Geographer Tim Cresswell (2004: 8) discusses how Tuan likens space to movement, and place to pauses – stops along the way – and, in this context, the workspace itself becomes a record of making and performance. According to Bosence, each stage of indigo dyeing – drawing the purple scum to one side, lowering the cloth, gently moving it in the liquid – is a movement to be appreciated: 'Think all the time, and the skill will develop. Later, you will be able to do this without thinking, and be free to talk, or think about something else' (Bosence 1985: 119). Tuan defines what is 'real' as 'the real, we feel'. It often goes unnoticed: 'Life is lived, not a pageant from which we stand aside and observe. The "real" is the familiar daily round, unobtrusive like breathing. The "real" involves our whole being, all our senses' (Tuan 1977: 145–6).

Bosence described the 'life' of the workshop as 'continuous', a place where 'imaginary tracks' are made around the space, determining routes and pathways to making in an area where multiple processes occur simultaneously 'so that no time [is] wasted waiting for cloth to dry, or a vat to mature' (1985: 77). She referred to these processes as 'operations' – a complex series of actions taking place in a managed, organized space kept 'clean and tidy, and spotlessly clean' (1985: 116). Bosence wrote that 'The plan of your workshop will be influenced by what you do', and described how a dyehouse and print room were planned at Dartington for teaching, a place to which 'members were able to come in between shopping and housework' (Bosence 1985: 75–7). The space was planned for 'our passage of work'. As dyeing is 'a very splashy business', paper work was done well away from the wet areas. Inside the vat room a concrete gully was built on two sides for hanging dyed cloth, but with the colours separated – indigo on one side, the other colours opposite. Later, at her home in Sigford, 'clean areas, dye gullies and hanging spaces were priorities', with sinks and taps located 'in convenient spaces', a mangle and low-level sinks for filling and emptying buckets and vats. Along one end of the workshop hanging lines were installed above a drainage gully – a 'ditch' with a soak away which was 'essential to catch surplus dye and drips'. The vats were placed in front of the lines so that 'as you lift the wet cloth out of the vat . . . all you need is a third arm to take one end of the cloth and hold it out . . . while the other two cope with the pegging!' (Bosence 1985: 75–7). Tuan observes:

> Most movements complete a more or less circular path, or swing back and forth like a pendulum. In the home . . . a desk, an armchair, the kitchen sink . . . are points along a complex path of movement that is followed day after day. These points are places, centres for organising worlds. As a result of habitual use, the path itself acquires a density of meaning and a stability that are characteristic

> traits of place. The path and the pauses along it together constitute a larger place . . . desk and kitchen sink are themselves important places connected by an intricate path, pauses in movement, markers in routine and circular time.
>
> Tuan 1977: 180–2

Tuan invites us to reflect on the routine of work. It is not celebrated – it is just another day, and yet rituals surround these 'to-an-fros'. Each day is a new day. In the morning the work lies ahead, in the future. The work may be dull most of the time, but some form of novelty or discovery is always a possibility. These characteristics – uncertainty, and the potential for surprise – contribute to a sense of future. At the end of the working day, the home becomes the future. The return journey is a tracing of steps back in space, going back in time to 'the familiar haven of home'. Tuan perceives familiarity as a characteristic of the past. The familiarity of home provides an image of the past. Moreover, home lies at the centre of life, signifying origin and beginning (Tuan 1977: 127–8).

From domestic workshop to professional studio

From this point, this chapter considers the transition from the domestic workshop to professional studio, and the influence of scale and volume on the making of work. When Barron and Larcher received important commissions from the Duke of Westminster, Girton College, Cambridge and Winchester Cathedral, alongside the production of regular orders from Muriel Rose at the Little Gallery, these large-scale projects became more challenging. The work demanded new ways of thinking. Barron and Larcher conceded that some jobs could be shared with others – scouring, washing, ironing, marking out the cloth and fixing the print in the steamer. They took on apprentices, teaching girls from the locality, and expanding their workforce to achieve their goals. Enid Marx began her design career as an apprentice to Barron and Larcher in Hampstead and she described her recollections of that experience:

> The work fell into a pretty regular routine . . . there were four of us printing . . . not all at the same time as there was much to be done: dye mixing . . . all its preparations . . . mordanting, steaming and washing. The last was an onerous job . . . when everything had to be washed and rinsed by hand, much of it hosed down out of doors. Then there was ironing galore.
>
> Marx 1985: 131

But Marx went on to say that, although much of the equipment was homemade 'of the "make do and mend" variety' (for example, the steamer had been converted from a metal dustbin), 'standards were professional and comparable to any manufacturer' (Marx 1985: 131). Design historian Alan Powers notes that Marx's experience of working in a craft workshop with a limited palette equipped her to work creatively during the Second World War, unhampered by the wartime shortage of colours that followed: 'From the small scale operation of block printing came the experience and inspiration to influence mass production', where Marx applied her skill as a designer to industry – for London Transport, and as an adviser to the Utility Furniture Committee (Powers 1992: 43).

It is significant that, although some of Barron and Larcher's products included bespoke garments and soft furnishings for domestic interiors, their craft production shifted from small-scale items to continuous lengths of dyed cloth and repeating pattern. Barron's understanding of indigo was gleaned from industrial chemists and large-scale manufacturers whose techniques she went on to improvise in craft production. She said, 'I was thirsting for an indigo vat' and, when she and Larcher moved to a house in Painswick, Barron turned the stable into a dyehouse:

> That kind manager [at Turnbull and Stockdale in Accrington] wrote down for us exactly how the vat must be made in the floor. When I showed the builder the plan he said, 'Oh, it's just a cesspit – quite easy', and did it very nicely for us [and] we had a railway in the ceiling for the star frames to run up and down, such as you have for a garage door.
>
> 1964/2018: 70

Barron and Larcher were selling cloth by the yard at a time when manufacturers were producing fabric on the roll. But most importantly, Marx stressed that manufacturers 'took the line that just *because* we were working on a comparatively small scale, we were in the position to break new ground in a way that factories with huge overheads could not afford to risk' (Marx 1985: 131). While this is testament to the innovations of Barron and Larcher and their contribution to English craft, their work is an example of the influence of rhythm and motion in making. Their practices illustrate a series of multiple movements in the studio, 'a meshwork of interwoven trails' (Ingold 2006: 50), each procedure following its individual pathway through the workspace while operating in tandem with another. Tuan observes that 'time and space are directed when one is actively planning. Plans have goals . . . and plans need not be grand to add a temporal dimension to oriented space' (Tuan 1977: 128).

Material flow: paste and wax

Bosence is recognized primarily as a printer, but she wrote, ‘For my part, I am happy to work in a resist method’ (1985: 90). Bosence developed a unique process of painted paste resist – a series of simple stripes that bely the mastery of her technique. Using ordinary household flour and water, she mixed a ‘pancakey consistency and tried to draw lines on calico’. The sticky paste was intractable. Tiny fibres on the calico were an obstacle to painting clean lines: ‘I should have starched the cloth first – the calico was ruffled and distorted.’ When she tried to pull the cloth into shape, she ‘noticed tiny scissions and rippling lines in the paste. Supposing dye could be pushed through them?’ Refining her techniques and consistencies, Bosence developed her application of paste and dye on starched cloth, pulling the hardened paste to reveal a lacery of lines and cracks in stripes and checks. According to O’Connor (1982: 44), by noticing what happened on the cloth Bosence developed a very personal use of flour paste resist as seen in her *Resist Patterned Length* (Bosence 1983). Although flour was

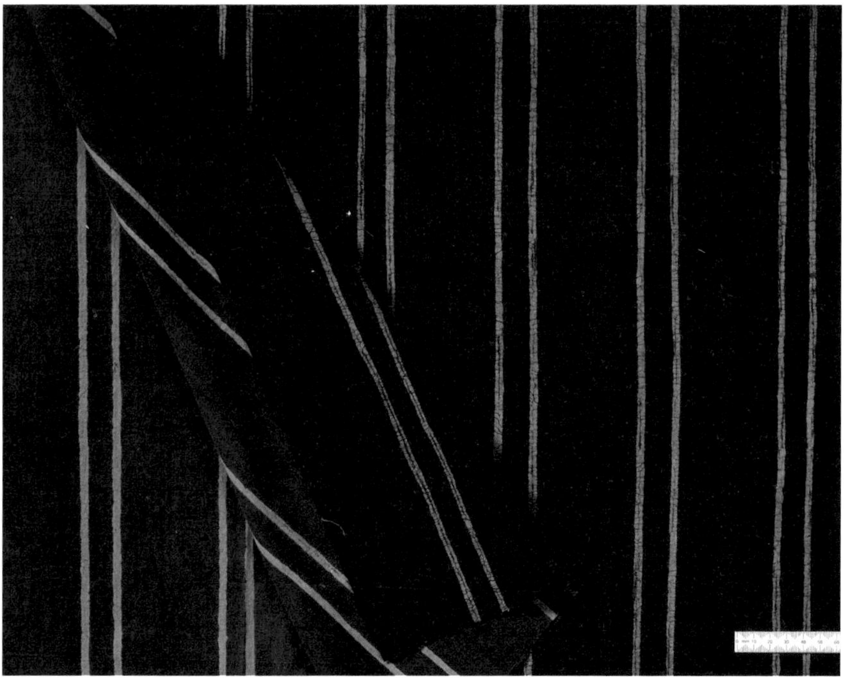

Figure 2.1 Susan Bosence (1983), Resist patterned length. Vat dye over resist paste on indigo dyed cotton, in The Textiles Collection at the University for the Creative Arts at Farnham, 7943. Reproduced with the permission of the University for the Creative Arts.

not sticky enough to withstand dyeing in indigo, O'Connor recognizes that Bosence used it underneath dye thickened for printing: 'When dry, the paste wrinkles up the cloth; stretched out again, there are fine cracks, which in her controlled hands, can give subtle veins of colour . . . Much sticky scraping and many washings are necessary here. The technique needs great control.' In O'Connor's view, '[t]he weight of cloth, combined with Susan's elegant sense of interval, make these fabrics work particularly well as tablecloths', and Bosence designed checked and striped tablecloths to complement the ceramic tableware of Ray Finch, Lucie Rie and Andrew McGarva (O'Connor 1982: 44).

Following in the footsteps of Barron and Larcher, Bosence received a commission for flour paste resist work on Irish linen as a set of curtains for Exeter County Hall (Bosence 1985: 124):

> This had to be planned as carefully as a military exercise. The material had to be scoured, dried, ironed, cut into curtain lengths, pasted and brush dyed within a given time . . . [Then] the flour resist paste, the thickened dye, the cloth, the area to work in and a large bath of sulphuric acid for fixing the colour [were prepared]. Two . . . expert 'pasters' assisted with the lines, another made up fresh paste as we used it. The cloths were resisted, dyed, dried and somehow . . . we managed the sulphuric acid bath. They were then washed, rinsed, dried and ironed, and sent off to be made into curtains.
>
> Bosence 1985: 124

Tuan reminds us that while people may crowd us, they can also 'enlarge our world'. In his view, 'When people work together for a common cause, one man does not deprive the other of space.' Rather, he increases it for his colleague by giving him support: 'Out of a crowded room a haven of warmth and tolerance is created' (Tuan 1977: 64–5). Bosence's account demonstrates that she acknowledged her need for support in the process of larger-scale production, thus making the shift from domestic practice to professional studio, but she admitted, 'I don't know if we would have welcomed a repeat order' (1985: 124). The work had been a carefully planned sequence of operations in its programming and completion, a sequence that Ingold described as following the forces that bring the object into being, aligning movements and gestures (Ingold 2009a).

In contrast to printing blocks, molten wax is the resistant substance that most accurately exemplifies Ingold's notion of 'material flow' (2009a). Its linear movement is dependent on the application of hot beeswax from a metal printing block (*tjap*), or through a *tjanting* – a short-handled 'pen'-like tool with a metal 'inkwell' at the tip of its bamboo holder. The wax passes through a fine spout, controlled by its temperature and the side-to-side movement of the hand. A subtle up and down movement of the wrist controls the angle of the *tjanting*, the flow of wax onto the cloth, the thickness of the line and weight of mark. These refined movements expand the potential for directional line, fluid trails, repetitive mark making and the clustering of small-scale elements *en masse* – a quality that O'Connor refers to as 'a wonderful ground' (O'Connor 1985b). In O'Connor's view, Bosence's work in wax is characterized by 'precision of placing and exact control of a fluid medium' (1982: 48). She associates Bosence's sensitive design of repeating spots with a simple, musical quality:

> They seem to ring clearly but softly like the different gongs in Javanese gamelan music. Each spot has an individuality because the wax, although controlled, does not have the cloned sameness of a mechanically produced mark, and the spacing and depth of tone give rhythm and resonance.
>
> O'Connor 1982: 48

Pattern, rhythm and repeat

Barron and Larcher (Silver and Burns 2018), Marx and Bosence belong to a group of textile printers of the twentieth century who introduced the simple, cut linoleum printing block to their repertoire of tools. Unlike the larger and heavier wooden blocks used to apply resist paste in Central Europe, hand-cut units were smaller, and lighter to hold and manipulate across the cloth. But, as Barron said, 'a design isn't dozens of little objects, or . . . rectangular window panes, it is something that becomes a design by repeating . . . It took me many days to make the design repeat at all. Finally, to give it more flow, I printed it upside down' (Barron 1964/2018: 56). The processes of Barron and Larcher and Marx explored combinations of direct printing, and they all employed discharge printing to remove colour – a chemical printing process that, when activated by steam, strips out dye to a white or light ground in printed areas. Although resist printing was not part of their practice, much of their work combined direct and discharge printing together with indigo dyeing. The principles of design and production in the work of these modern textile printers is mirrored in the formation of resist dyeing in all its variants. In printed textile design, pattern flows out from its edges beyond the structure of its geometry. Whether direct, discharge or resist printed, pattern interlinks with itself to create fluid, uninterrupted composition over the surface of the cloth.

Textile designer Neisha Crosland comments that the strength of Barron and Larcher's designs is in their 'energy', the 'momentum in the sequence of the repeated motif'. Importantly, Crosland remarks, 'the unprinted areas of plain calico cloth played a part in defining the design, and were as important as the printed spaces' (Burns, Crosland and Loakes 2018: 101–4). Crosland's observations demonstrate the positive and negative space critical to placement and design. Similarly, in his survey of *Modern Block Printed Textiles* (1992), Powers describes how Marx rotated the block in alternating upside-down prints to create 'subtly organised pattern repeats'. Some are very simple, others with widely-spaced stars or dots. But they are 'never quite regularly spaced, nor repeated with quite the same intensity, giving the whole design a sparkle and humour'. Powers observes, 'It is clear that [the design] has grown under the hand and eye of its own creator, rather than being laid out on a grid by a machine' (Powers 1992: 42).

Bosence is noted for her rhythmic block prints, often overprinted to create dense textural effects. She emphasized the power of movement in designing, cutting and printing her blocks. In describing her cutting tools – knife, chisel, graver –Bosence drew an analogy with a golfer and his clubs: 'His irons and woods and putters help him hit a small ball for short distances, to manoeuvre it out of a ditch, and to coax it into a hole' (1985: 101). When printing the block, any directional leaning in the left- or right-handed person can influence the course of the print. Guidelines and registration points are marked on the cloth but 'it is difficult not to let the block "walk" a fraction if always working from the same direction' (Bosence 1985: 102). She added, 'Whilst working . . . ideas come from everywhere, all the time – printing in different ways, resisting parts, overprinting, variations and adaptations. Make notes . . . one has to catch them quickly' (1985: 103). Consequently, it is clear that an intrinsic understanding of pattern and repeat is central to the organization of any print workshop, and its principles reside at the heart of a textile artist's process of thinking.

Rhythms of resist: 'blueprint'

Printing with lino blocks echoes the rhythm of resist printing at the indigo workshop of Stanislav Trnka in the 1980s. In Púchov, Slovakia, Trnka's 'blueprint' process was dependent on the even distribution of resist paste through a succession of rhythmical movements between tiered tray, printing block and cloth. Three or four light dabs of the block were sufficient to load it with paste. Its sharp, fluorescent green revealed a concoction of heavy metals in the mix. When dry, this refined paste was unusually resistant, and held fast to the fibres of the cloth through every stage of soaking, dyeing, stretching, reprinting and overdyeing. Sharp pins at the corner of the block enabled accurate registration, and Trnka's printing proceeded with speed and perfect rhythm. As the images multiplied, the printed cloth was taken up and over rollers while fresh, unprinted sections were shifted along the print table. The printed cloth was wrapped concertina-style around the hanging rack of nine rails suspended from the ceiling. The frequency required for reloading the tiered tray (the 'sieve') with the paste (the 'pap') varied from one block to another, sometimes after as little as four prints of solid pattern, to as many as twelve prints of a finer image depending on the density of printing surface (Brassington 1987: 34–9).

The skill was in the smooth repetition of the image, the uniformity of the resist paste, the regularity of the print; demonstrations of a tacit knowledge through orchestrated movement. Personal experiences of printing in Púchov under Trnka's watchful eye are recalled in the following passage. The use of italics denotes the experiential voice – the memories recollected. Taking the first tentative steps to the print table, and holding the old wooden block with the 'sieve' of 'pap' alongside, the process was set in motion:

> *I hear the sound of surface tension as the paste is picked up, the strike that transfers image to cloth. Guided by the registration points at each corner of the block, I place the nearest left-hand pin on the central registration line, using my left hand to steady it. I lower the right-hand pin on the line, allowing the block to gently drop into position and, with a few light taps of my fist in the centre, and at each corner, the 'pap' is transferred to the cloth. 'Don't lift the block too quickly'. The tacky surface drags the cloth with it and the sudden movement flicks small splashes across the pure white ground. The printing begins at the centre and continues from left to right; working the second and subsequent rows away from me towards the opposite selvedge. Returning the block to the centre, the pattern is continued in the opposite direction; row by row towards me, avoiding the need to lean over the wet print. 'Wipe the pins clean with your fingers'; a necessary process to prevent an accumulation of dots from interrupting the repeat. My arm is aching. The block feels heavier with every movement, more difficult to manoeuvre between the 'sieve' and the cloth, more cumbersome to support when registering the repeat. This is hard, physical work.*

Tuan emphasizes the importance of the forward motion: 'Normally a person feels comfortable and natural only when he steps forward. Stepping back feels awkward' (1977: 129). In printing, the block, and the ideas that follow it, move in all directions – forwards, backwards and from side to side, shifting left or right into a 'brick' formation, or downwards to create a 'half drop' placement. It can be seen, therefore, that the influence of pattern and repeat in the motion of the workshop is significant in the performance meanings of resist printed cloth.

Early experiences of resist block printing in Slovakia prompted a work of regular lines and dots on 5 metres of linen (*Crafts Study Centre Collections*, ID Number Current Accession 2002.33). Rather than printed, the *Resist dyed length* was waxed (Brassington 1985). The markings were applied in layers, using a smaller Javanese *tjanting* and a brush, and then overdyed in indigo, iron and tannin in successive stages. The work consists of hand-drawn wax in continuous, moving line. The *tjanting* was passed from left to right and from right to left in

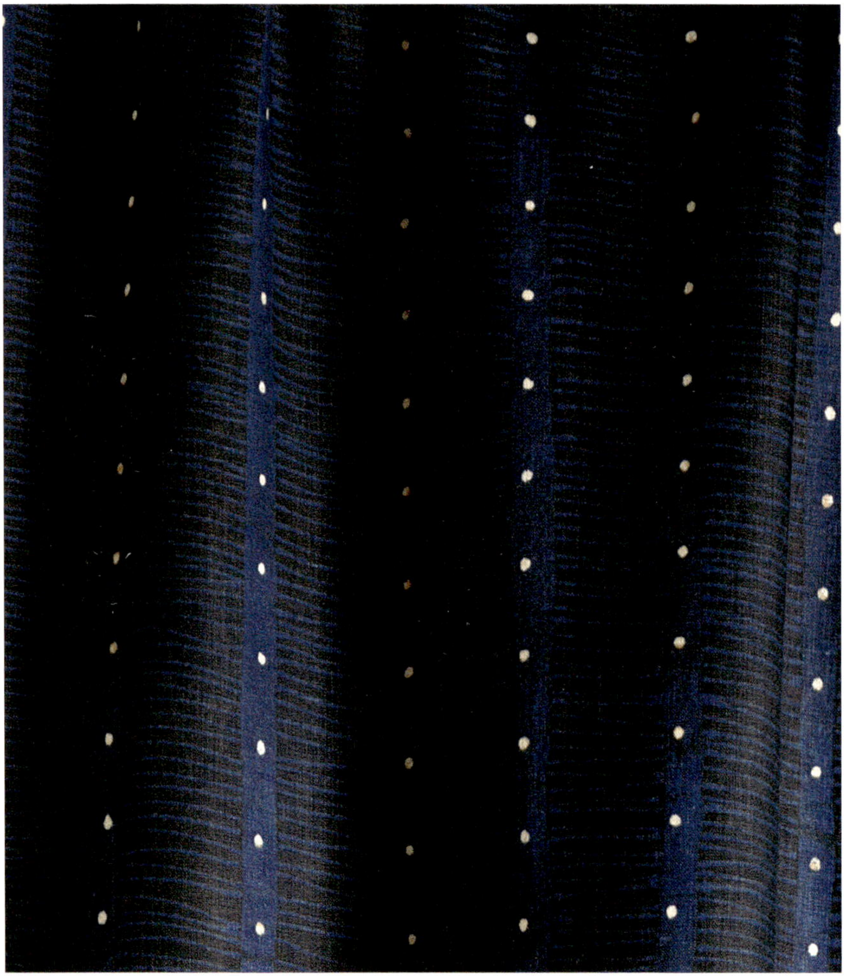

Figure 2.2 Linda Brassington (*c.* 1985). Waxed length. Linen, dark blue indigo background with bands of lighter blue and cream spots. Hand drawn wax resist. Spots waxed and overdyed in indigo. Lines waxed and overdyed in myrabolan and iron rust, Crafts Study Centre, 2002.33. Reproduced with the permission of the Crafts Study Centre.

narrow sections down the cloth, and an accumulation of fine spots was dotted downwards in a sequence of stripes – as 'stops along the way' (Cresswell 2004: 8). The piece was made in direct response to the rhythm of 'blueprint' (*modrotlač*) in Slovakia. It is a reinterpretation of systematic print, expressed in the flow and movement of hot wax. Most importantly, irrespective of technique, the motion in its making echoes the principles of repeat and repetition embedded in textile practice.

Rhythms of *shibori*: cultural connections

In *The Textiles Collection* at the University for the Creative Arts at Farnham, one particular artefact (ID Number Current Accession 8090) seems to express a common rhythm and flow in the textile practices of England and Japan. In *Hand*

Figure 2.3 Japanese shibori. Vertical irregular bands of alternate soft blue, dark blue thin moving lines, *c.* early twentieth century. Indigo over hand spun and hand woven cotton from the source collection of Barron and Larcher, The Textiles Collection, University for the Creative Arts at Farnham, 8090. Reproduced with the permission of the University for the Creative Arts.

Block Printing and Resist Dyeing (1985), Bosence recounts that the piece was found by Barron 'packed around a tin of dyestuff from the East'. Bosence describes it as 'Japanese fine cotton, which was bound and tied into a long snake, then dyed', and she notes the fineness of the pleating and binding as 'exceptional' (Bosence 1985: 59). Now, the piece carries a handwritten note pinned to its edge: 'Tie dye cotton. There were about 14yds of this narrow material, tied in a long sausage.'

This discarded packaging was retrieved, recycled into clothing and absorbed into Barron and Larcher's life, seeming to yield to a new context and identity. The extraordinary encounter with this strip of indigo dyed cloth seems to exemplify a cultural connection. Here, the rhythms of English printed textiles and the flow of Japanese *shibori* wrappings seem bound together. Bosence cited an article on Japanese *shibori* published in *Time* magazine in 1976: 'Only one craftsman could work on the length of cloth, since even a change of workman – or even a brief illness – could result in an irreparable alteration to the rhythm of the tying and the evenness of the results' (Bosence 1985: 60). This *shibori* cloth and the reference from *Time* magazine reflect the cultural connection between them.

In their extensive examination of *Shibori: The Inventive Art of Japanese Shaped Resist Dyeing* (1985), Yoshiko Wada, Mary Kellogg Rice and Jane Barton outline the development of the craft in Arimatsu. The town established itself as a centre of *shibori* production in Japan during the seventeenth century (1985: 28). It developed its resist dyeing processes to capitalize on the passing trade of feudal lords, merchants and pilgrims on the Tokaido route travelling from Kyoto to Edo (now Tokyo). The town's production was based on locally grown indigo and cotton. While early patterns were variations on wrapped resist dyeing, the increasing market demanded new approaches, and generated innovative developments in knotting, pleating and binding to achieve a broad range of fine markings. Wada, Kellogg Rice and Barton explain that a new method of looped binding was introduced by a doctor's wife, Mrs Miura, who had travelled on the Tokaido route but eventually settled in Arimatsu. They describe the *miura* technique as incorporating a fixed hook that supports each small section of cloth while the thread is wound and the knot is tensioned and tied. The fixed hook, as well as speeding up the process of binding, also enables the cloth to be tensioned into sharp pleats, resulting in fine, resist dyed lines and folds (1985: 28). As the variety of patterning grew, and sales increased, a division of labour was introduced that was comparable with the developments in English craft practices. According to Wada, Kellogg Rice and Barton, 'There were those who drew the designs, bound or stitched or pleated the cloth; those who dyed it, and merchants who sold it' (1985: 29). In Tuan's view, 'Tools and machines enlarge man's sense of space and spaciousness', feeling them to be a 'direct extension of his corporeal powers' (1977: 53).

Production thrived in Arimatsu for more than 300 years until political change in the early twentieth century brought increasing industrialization, and economic decline threatened the production of *shibori* in Arimatsu (Wada, Kellogg Rice and Barton 1985:33). In response, local resident and inventor Suzuki Kanezo set about finding:

> . . . a new and faster method of dye patterning cloth . . . devising an ingenious way of creating patterns by wrapping the cloth around long poles, pushing it along the pole into tightly compressed folds, and immersing it, pole and all, in a long vat of indigo . . . Production increased at least tenfold . . . [enabling] two workers to produce twenty bolts in a day . . . competitive with printed cloth, but [with] an entirely new look.
>
> Wada, Kellogg Rice and Barton 1985: 34

The pole-wrapping process resulted in the 'poetically named *arashi*' meaning 'storm' (Wada, Kellogg Rice and Barton 1985: 34). During the years of prosperity in the 1920s and 30s, *shibori* was adapted to Western-style clothing. From the perspective of Wada, Kellogg Rice and Barton, this use of *shibori* was 'aesthetically unfortunate' but was 'a symptom of vigour rather than of decadence' (1985: 35). At the same time, chemical dyes were introduced and new devices followed. Folding and clamping cloth between boards (*itajime*) gave rise to large-scale repeating shapes. Resist dyed in bright yellows, reds and indigo, these cloths became a popular export for the African market (1985: 34).

Textile curator Matilda McQuaid places the practice of the late Hiroyuki Shindo in the context of Japan's textile history: 'No other colour is more associated with Japan than the deep blue of *ai*, indigo, and it was not unusual for villages to have a specialized indigo dyer to meet the demands of the local production.' Shindo, 'enraptured' by the colours of natural indigo, was 'saddened' by the decline in indigo growers, and resolved to revive and 'transform' the craft (McQuaid in McCarty and McQuaid 1998: 21). Shindo processed his indigo leaves with the addition of lye, lime, wheat bran, sake and a 'microorganism' to aid fermentation. McQuaid notes that Shindo's lye came from ash accumulated in his woodburning stoves, and the remaining ash and spent indigo leaves were later used to fertilize his garden. In McQuaid's view, 'Shindo's art and life remain inextricably linked to nature' (McQuaid in McCarty and McQuaid 1998: 21).

> Shindo explains his space dyeing [as] *okkochi* meaning 'eastern wind' which, when spoken, also suggests 'to let fall'. [It] describes a kind of dyeing technique developed in Arimatsu and the story of its origin claims that a wind from the east blew just a corner of kimono fabric into a vat of indigo. From this incident, a new shaped dyeing was born that was less methodical . . . and more reflective of the patterns of nature.
>
> McQuaid in McCarty and McQuaid 1998: 21

Shindo worked with a collection of eight indigo vats sited at the heart of the workshop. During field study visits in 2007, 2019 and 2024 the same vats continued to stand as iconic monuments to his practice. Organized in rows, with flexibility to move around and across them in the process of dyeing, Shindo's indigo vats are the site of motion, making and gesture. In contrast, his workspace is situated alongside a raised platform – a place of stillness for the presentation of work, an interface between studio and gallery. Similarly to Bosence, the transitions between Shindo's home, garden, studio and gallery are seamless.

In comparison, textile artist Shihoko Fukumoto's 'flow' of work is dependent on climatic conditions – the air, the weather, heat and humidity (Millar 2001: 49). Fukumoto activates a succession of pulleys above her indigo dye vats, and sprays a series of horizontal waves of water across her cloth, repeatedly dyeing and washing each length over a period of four to five days. Curator Mary Jane Jacob views Fukumoto's work as embracing the concept of nature and time, an unhurried commitment to a deep understanding of process: 'Fukumoto has exchanged the tie dye process of *shibori* for the applied dye process of *bokashi*' (Jacob 1993: 10–11). Her operation of repeatedly dipping and raising the fabric from the vat, and blending and moving the threads of her loosely woven cloth with a brush, forms rippling waves of linear movement. Jacob describes the results as 'pure tonal variations whose fluid, wavy lines seem to be evidence of the liquid indigo dye' (1993: 10–11). Fukumoto's process is the essence of Ingold's 'fields of force and flows of material' (Ingold 2009b: 91). Her rhythmical motion of dyeing, spraying, brushing and blending traces the movement of indigo along the threads of her cloth, 'following the forces that bring the object into being, aligning movements and gestures' (Ingold 2009b: 97). Fukumoto's smaller series of *bokashi* are interpretations of historic paintings and calligraphy scrolls incorporating gold leaf. In these works, both the supporting paper and the cloth mount are dyed in indigo. Rice paste is the counterpart to hot wax but the substance is applied cold through a stencil (*katazome*). Fukumoto applies *katazome* paste resist and gold leaf before overdyeing the whole composition in indigo to achieve patches of luminous emerald green – a material mix of paper, cloth and gold in abstract composition.

Rhythms of Mashiko: a personal narrative

According to stencil dyer Eisha Nakano, 'the Japanese have developed *katazome* to a degree of sophistication unmatched elsewhere' and 'no other culture can match the level of dye expertise attained in Japan, a country that has nurtured a tradition of hand dyed textiles for hundreds of years' (Nakano 1982: 3). From this point, the chapter presents a personal narrative of *katazome* and indigo dyeing at the historic *Higeta Aizome Kobo* (Higeta Indigo Workshop) in Mashiko, Japan. It is the family workshop of Tadashi Higeta, whose father was a master craftsman in *katazome*. Higeta's two large volumes of stencils reveal the exquisite quality of their detail. Inside the covers, minutely fine marks seem to dash and dart over the pages, creating beautiful rhythms on dark, crisp paper. The deepening tones of persimmon juice (applied to create a waterproof surface) have intensified over the decades, and have aged the cut stencils to a strong, glossy, bronze black. Small dots and lines resemble seeds blown in the wind, ripples on water.

The following account records the personal experiences of an art residency at the *Higeta Aizome Kobo* in 2016. It reflects on the significance of the space; its influence on movement and motion in the making of new work over a period of five weeks. The passages follow a 'path traced through the terrain of lived experience; a meshwork of interwoven trails' (Ingold 2006: 30):

> *The dye room is filled with an expanse of seventy-two indigo vats buried into a raised platform of earth. It is situated on one of four sides of a quadrangle, where the flow of multiple practices cross and circulate in its central courtyard. Opposite the indigo workshop, I enter a dusty room piled high with paper, cloth and tools above the earth floor. Takumi Kojima pulls out a panel of plywood and slings it across a pair of trestles. Taped above the shelves are more sheets of plywood, as flaps to keep out the dirt. They are lifted and a sample of cloth is dragged out; a narrow, woven, textured stripe on cotton. The board is sprayed with water to release a thin layer of rice paste, and the cloth is smoothed flat across its surface and stuck down. A large pot of 'glue' is uncovered beneath its clear polythene skin, and a lump is scooped out onto a mixing board. Rice bran and white chalk are added, and the ingredients firmly mixed with a wooden spatula. The paste is a fine balance between alkalinity and acidity, judged by the depth and sharpness of its yellow colour. My tentative applications begin. I draw small squeegees up and down each section of the cut image. I try to build a thin, even layer thick enough to resist the dye while retaining a clear edge to the design. Then, I squeeze a trail of it through a piping bag. The technique is difficult and demanding. This goo is thick, heavy and*

glutinous. It seems almost inflexible. 'Am I pressing too hard? Am I scraping it away rather than applying it?' As I spread it through the stencil and onto the cloth, this is not like the pressure of screen printing. This is about gliding the paste over the surface, and yet it seems far too stiff to glide. Piping it through a cone to which metal nozzles are attached – a small hole, a larger hole, a slit – bears no resemblance to sugar icing. This mix is thick and requires muscle to force it down through the bag and out through the opening. Once the paste hits the cloth it is reluctant to attach itself to the fibre. Instead, it appears to lift off the surface as it is released from the cone. This is extraordinarily difficult to control.

Resist paste is a challenging but intriguing material. It soon became clear that the traditional tools – spatulas (*hera*) and piping cones (*tsutsu*) – determine the marks and gestures on cloth. When using a *squeegee* – a tool with a thin rubber blade – there is a tendency to push vertically, up and down. In contrast, the curved handle of the wooden *hera* encourages circular movements, forming expansive arcs across the surface. The wet 'drawing' was sprinkled with a drift of fine woodchip, confining the paste within the edges of the design, and the excess was brushed away. The stencil was lifted from the cloth in a single, positive action, and promptly washed. When these humble attempts were complete, a bamboo stick was pushed under the cloth to lift it from the board. Each end was cleverly attached to a series of pins on pairs of wooden strips (*harite*), and the cloth was clamped at both

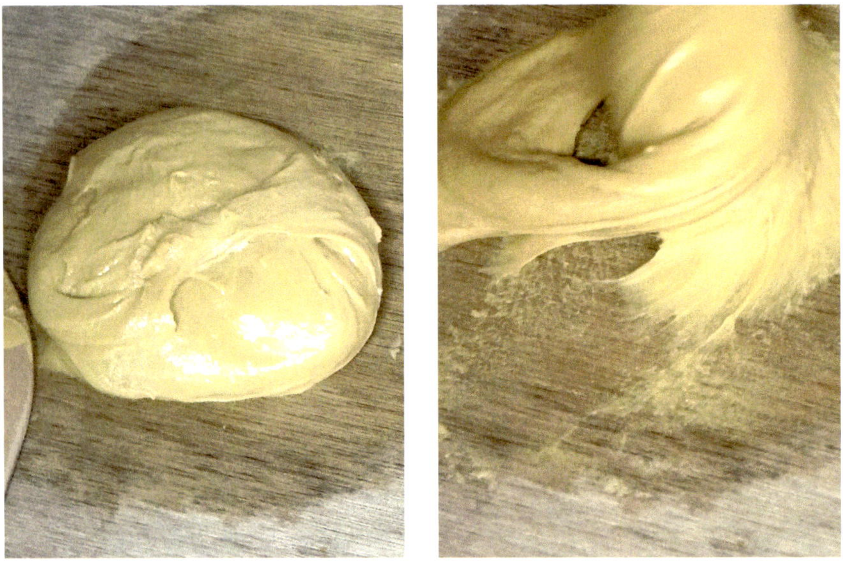

Figure 2.4–2.5 The mixing of *katazome* resist paste at *Higeta Aizome Kobo* in Mashiko 2016. Photo: Linda Brassington.

Figure 2.6 The author applying *katazome* resist paste on linen and *sasawashi* paper cloth at *Higeta Aizome Kobo* in Mashiko 2016. Photo: Mayumi Tsukimura.

edges. Loops of fine rope were attached, tensioning the cloth, and flipping it over. The taut cloth was sprayed with water and vigorously smoothed with a flat wooden stick, backwards and forwards, from side to side across the cloth, pushing the moisture into the 'glue' and forcing the sticky substance into the fibres – but from the reverse. This was a revelation – demonstrating how resistance to the dye is ensured on *both* sides. The cloth was carefully pinned and suspended from a drying frame. It was transferred to a heated cabinet where it set into a hard textured relief – a surprisingly resilient crust preventing dye from penetrating the stencilled areas and prompting new observations:

> *But this thick, tactile coating offers other visually rich qualities, its opaque surface contrasting with the transparent layers of dye that build around it.*

As the sample strips were suspended in the makeshift drying cabinet, two 'Heath Robinson' electric fan heaters were switched on, their hoses projecting upwards, and the doors closed. Kojima calls:

> *'Come back next week when the glue will be dry.'*

These early attempts were uncontrolled and resembled an apologetic pastiche of Japanese line and shape. As the residency evolved, a new assemblage of pasted cloths was developed, and preparation for indigo dyeing began:

Every cloth must be passed through water, 'slowly' to avoid bubbles emerging and preventing the cloth from floating back up to the surface. The time is critical – just enough to thoroughly wet the cloth but to avoid the resist paste from breaking down. 'One or two minutes.' Each piece is swept up from the depths of the stainless steel bath, and suspended from a protruding wooden pole temporarily but effectively wedged under the window frame. Gradually the drips of water subside.

The cloths were hooked with safety pins along a strip of wood punched with holes. A simple loop was taken across the work and tied over a length of split bamboo. This supported each piece across the rim of the vat during dyeing, enabling it to be raised and rested across trestles between immersions as the colour oxidized. Each cloth was carried into the indigo workshop on its bamboo rod. L-shaped frames, like branches, were suspended from the rafters. The cloths were hooked over them, ready to be processed in succession:

Each piece of work is immersed in indigo and raised; the rapid contact with oxygen transforming the dye to a blue tinged with green. I tread a steady path over the vats to the light outside and plunge the cloth into the plastic trough of crystal-clear water. Should I stop, repeat, or wait and think again? For a deep, dark hue, repeated immersions of 'a minimum of ten seconds to three minutes' are interspersed with successive periods of oxidation of 'at least three minutes'. The darkest shade is achieved with ten or more immersions, oxidizing the cloth for as long as possible in between to retain the strength of colour. Can I also achieve a pale, transparent tone that is subtle, but fast and even? 'Three times in indigo, but just for ten or twenty seconds. Take the cloth outside to the water, rinse it, keep checking the colour against the sunlight. Are you happy with it?' The cloth is returned to the sinks, plunged into another trough of water and left overnight to gently dissolve away the paste. This is critical. 'If the glue is removed too rapidly a stain is left behind.' As the evening draws in, I examine the wet cloths under water in the diminishing light. It is dusk. Every day, there was a sequence of movement across the inner courtyard – a rotation from the sinks to the indigo vats, to the large plastic trough of cold water precariously balanced beneath the tap, to the clothes line suspended from a frame of wooden poles.

Slowly the work emerged. Each composition was considered in relation to another; making judgements and readjustments through a series of parallel explorations on cloth; ideas in clusters, responses in pairs and groups. *Katazome* paste defined and maintained shape with surprising clarity and consistency, but its powers of resist revealed subtle variations on different cloths; sometimes blocking colour from both sides, sometimes revealing partial penetrations of dye. These qualities were explored on different weights and structures, weaves and substances; single and double-sided, smooth and textured. Each piece was turned and reversed; an interchangeable series, developing shifting rhythms and sequences from front to back. Higeta observed the process, the accumulation of colour and tone on cloth, the results as the excess dye was rinsed away. One by one, the colour and composition were revealed on each piece. Gradually, as the work materialized, new layers were envisaged. Confidence developed – discharging, dyeing, stencilling, overdyeing – to build a series of compositions on different cloths:

> *Each day, I walk a rotation from the outhouse, into the smoking atmosphere of fermenting indigo, and outside, over the muddy, well-trodden pathway to the washing lines; a repetition of processes and flows of movement over minutes, hours, days; a choreography of craft.*

Rhythm (Brassington 2017) was produced on return to the UK from Mashiko. It consists of rhythmical bands of indigo down a 5-metre length of narrow cloth gifted by Tadashi Higeta. At Farnham the work unfolded through a sequence of processes in the studio. The cloth was stretched, tensioned and masked with a single continuous stripe of discharge paste down its centre. Peeling the dried cloth away from its sticky residue, it was lifted and carried to the tall steamer that accommodated its length in one continuous suspended spiral. Heat and humidity activated the dried paste and stripped the natural colour from the linen fibres, 'bleaching' it to white. After washing, drying and ironing, the cloth was repositioned on the print table and masked with a second layer of repeating horizontal bands, spreading *katazome* paste from selvedge to selvedge to set hard against the surface. Flipping the cloth, the application of pasted stripes was repeated on its reverse. A sequence of rhythmical movements across the cloth, a pathway from print room to steam room – steaming, washing, drying, ironing – echoed the rotation of movement in Mashiko and marked the passage of time through the studio space. Tentatively, the resist printed cloth was passed through a weak solution of indigo in a wide shallow trough, and watched closely. As the dye began to reveal shadows of transparent blue, the cloth was swiftly removed, catching the point where the paste began to break down. Weeks later, under the gallery spotlight of the Mashiko Museum of Ceramic Art, *Rhythm* became a configuration of movement and motion in making, the expression of 'a path traced through the terrain of lived experience; a meshwork of interwoven trails' (Ingold 2006: 30).

Figure 2.7–2.8 Stencils suspended from a drying rack in the wash room at *Higeta Aizome Kobo*, Mashiko, Japan, 2016. Photo: Linda Brassington.

Figure 2.7–2.8 (*Continued*).

Figure 2.9–2.10 Dyed paper suspended above the indigo vats to oxidize at *Higeta Aizome Kobo*, Mashiko, Japan, 2016. Photo: Linda Brassington.

Figure 2.9–2.10 (*Continued*).

Figure 2.11–2.14 Movement and motion through the processes of making *Rhythm* in 2017, applying discharge and indigo over masking tape and *katazome* paste resist on linen. Photo: Linda Brassington.

Figure 2.15 Linda Brassington (2017). *Rhythm* exhibited in *Linda Brassington: Work from the Art Residency 2016* at Mashiko Museum of Ceramic Art, 2017. Photo: Linda Brassington.

Figure 2.16 Linda Brassington (2017). *Rhythm*. Indigo over discharge and *katazome* paste resist on linen. Photo: Linda Brassington.

3

Material as Metaphor: Colour as Stuff

This chapter explores the boundaries between dye and pigment, depth and surface. Reflecting on the theories of philosopher John Ruskin and anthropologist Tim Ingold, the chapter relates the stuff of colour to the performativity of its processes. It introduces indigo as material, a living substance that changes and shifts in its process of transformation; a phenomenon in which 'fields of force and flows of material' (Ingold 2009a) form the making of work. The study begins with the investigations of key researchers in the field of indigo dyeing – Dominique Cardon, Jenny Balfour-Paul, Deryn O'Connor and Edward Bancroft. Their common thread is the recognition of indigo as a living, breathing substance, and the contemporary practices of artist Ryuta Sasaki are introduced to illustrate a deep, emotional connection with dye – indigo as a being, with a life from birth to death. As such, the chapter challenges the perception of indigo as merely 'blue'. Rather, the discussion reflects on the power of 'blueness' to express volume, physically and metaphorically. It considers how the physical and emotive responses to blue and blackness can generate a new understanding of indigo as an expanse of space – an all-enveloping darkness in its intensity of colour and depth of tone in two and three dimensions – identifying indigo with the resonance of darkness in human perception, its 'voice' and how it 'speaks'. In doing so, this chapter associates the stuff of colour with the material approaches of contemporary artists Anish Kapoor, Richard Long, Manish Nai and Chiyoko Tanaka, whose works exploit pigmented surfaces to visual and poetic effect. It reflects upon the relationship between a material and its source, and considers how location influences the application of pigment and the development of its practices. As a result, the chapter explores how expression with indigo and minerals (mud, iron and tannin) bears a direct relation to time and place, and the environment within which these works are created – the earth, nature, the ground beneath our feet.

Figure 3.1 Compacted earth surrounding an indigo vat at *Higeta Aizome Kobo* in Mashiko, Japan, 2016. Photo: Linda Brassington.

Indigo as substance

In Dominique Cardon's translation of an eighteenth-century dyer's records, the historian notes that the anonymous author begins his memoirs with these words: 'Blue being the primary colour, and the ground for the greatest part of fast colours, it is necessary to start with it' (Cardon 2016: 106). Any secondary or tertiary colours must be achieved by overdyeing after indigo. Other dyestuffs cannot be added to the indigo vat. Consequently, Cardon's description of the dye process emphasizes the significance of indigo as the 'primary colour':

> [Indigo] is deposited onto textile fibres by the precipitation of its soluble colourless form into its insoluble blue form under the action of the oxygen in the air. It forms microlayers on the fibres, the accumulation of which by successive dippings into the dye bath gives more and more saturated and dark blues. These progressive degrees of blues served as a kind of scale to dyers, who evaluated them by visual comparison with a standard of cloth swatches dyed in the successive shades.
>
> Cardon 2016: 105

Cardon notes that, historically, many shades of indigo have acquired names that have been handed down through the centuries – Persian blues, Turkish blues, azure blues, celestial blues. In her study of an eighteenth-century European scale of colour, Cardon observes that Persian blues were always at the top of the scale, where they had prevailed since the fifteenth century in the documents of Venice and Florence. At the other end of the scale, 'milky blue', *alatado*, figured as the lightest shade of blue in medieval Venetian cloth. In contrast, azure blues, *azuri*, and celestial blues, *celestri*, were classified in medieval Italy as darker than Turkish blues, while in the eighteenth century, in the Languedoc region of France, these blues were classified as distinctly lighter than Turkish blue. But Cardon declares, 'It cannot be overemphasised that these names are not merely poetical descriptions.' They became a visual standard, a colour comparison for dye and pigment (Cardon 2016: 106).

In contrast to the breadth of visual description for indigo blue, indigo dye is matter, a substance extracted from the fermentation of *Indigofera tinctoria* plants. The preparation of natural indigo for dyeing is an organic process – a performative procedure from the very beginning of its fermentation. In her comprehensive study of indigo, Jenny Balfour-Paul (1998: 104–5) records the first stage where leaves are cut, ground, pounded and moistened to achieve a composted sludge. She describes how dried leaves are crushed and smashed with mallets by 'rammers', the heap repeatedly sprayed with water as it heats, and the deep pile turned regularly to produce a fermenting mass of living matter. While Cardon chronicles the crushing of woad with sledgehammers in eighteenth-century France (2016: 77), woad being the precursor of indigo in Europe, Balfour-Paul describes the parallel process of *sukumo* in Japan, 'a huge steaming heap, containing up to three tons of leaf material'. She explains that the composting leaves of indigo, covered with thick straw referred to as 'bedcovers', are located in a fermenting chamber known as the 'bedroom'. They remain there for three months, after which the dense slime develops to a darkened paste and solidifies (Balfour-Paul 1998: 106–7).

Jin Kitamura is an indigo farmer in Matsuzaka, Mie Prefecture, Japan, who produces *sukumo* using traditional practices established centuries ago. In the film *Blue Comes from Indigo, Yet it is Bluer than Indigo Itself*, a Fantasia Yoihi Project 2024, Kitamura explains that *sukumo* breaks down through fermentation. The only variation available is adjustment to the height of the piles of leaves and the amount of water applied. Water is 'very, very important':

> We've tried different kinds of water. I realized that the water they used long ago was different from the water now. Now, we use water from the source in the mountains . . . We can't get the same water they used long ago, but water from the source is close to it.
>
> *Blue Comes from Indigo, Yet it is Bluer than Indigo Itself*, 2024

For Kitamura, it is critical that his farm produces *sukumo* without impurities. He says of growing indigo, 'I don't know about herbicides. I've never used them. I don't have that knowhow.' When removing insect larvae from the roots, he says, 'We do it by hand and it works', weeding around the plants and digging underneath any wilting plants to extract the offending larvae from the soil (*Blue Comes from Indigo, Yet it is Bluer than Indigo Itself*, 2024).

Cardon notes that this compost of leaves – the 'thick slur of indigo mud' – is a world practice. The resulting sediment, retrieved from the rotting organic matter, is drained. It develops into a damp paste that is dried to a precipitated 'cake', or moulded in the hands to form balls that are left to dry, and then stored (Cardon 2016: 105). Examples of dried, composted indigo balls are held in the Textiles Collection at the University for the Creative Arts. Handling them conveys the lightness of this dehydrated dyestuff, the impermanence of its material:

> *Cradling the crusty sphere in the palm of my hand, the dry, delicate form reveals its fragility. Its tenuous structure resonates with the vulnerability of its material. A light, crumbling surface of skeletal stems protrudes through the powdery pigment.*

Its tactile mass gives way to a dusting of purple particles on my hands, scattering its fragments into a fall of sooty substance onto the protective tissue paper beneath.

In this state, the dried and concentrated indigo becomes the key element of the dye vat in a process of renewed fermentation. In water, with the addition of an alkaline solution, the indigo is dissolved with the aid of a 'reducing' agent – a chemical that raises oxygen to the surface of the vat, while simultaneously drawing the indigo down into the depths of the vessel where it appears as a brilliant yellow-green liquor in the vat.

Deryn O'Connor witnessed a similar procedure in Guizhou Province, southwest China, where the stalk and leaves of indigo-bearing plants are harvested, soaked in barrels or troughs of water and left outside in the sun. She recalled that 'a stick is wedged across the top of the barrel to keep the leaves below the surface' and the mix is left to ferment for three days, by which time 'the leaves turn dark blue green and a dark scum appears on the surface of the water' (1994a: 69–72). In 1993, in a small green spiral-bound notebook, O'Connor had recorded her observations in a characteristic form of notetaking – capturing key points in a single line, breaking the lines into indented text, and using capital letters or underlining to stress a key element or a performative process. O'Connor's notes are a sequence of words – prose appearing as poetry:

October 2 Sat 1993
old lady
Grows in field about size of her courtyard
about c. 25' x 25' of STROBILANTHES flaccidifolius
in February takes cuttings & puts them
under plastic
When they crop it they pull up the
whole thing
pick them August/September
put in stone vat of leaves + LIME
60 gin of plants = 30 kilos
makes 2.5 buckets of paste a year
(Aug/Sept) (a gin = 0.5 a kilo)
Leaves and lime in for
3 days & nights and BEATS if
water drawn off (to get oxygen in)
vat in sunny part of courtyard
Keeps paste damp
Keeps dye vat going all year

> But 'FEEDS' it
> with a little paste
> + a little commercial
> sodium carbonate
> + Rice wine – helps the
> fermentation
> every night she feeds
>
> O'Connor 1993

Later, on 4 October 1994, O'Connor recorded seeing indigo in a vat. As she experienced the episode, she wrote about it in her diaries, capturing the performativity of the process: 'A girl came and got out [the indigo leaves], and sieved [the liquor] before adding lime from a large pudding bowl.' According to O'Connor, the vat 'slowly went milky' and the girl 'moved it round carefully with her hands – gently – using a gourd like a handled dish' (O'Connor 1994b). In her publication *Miao Costumes* (1994a), O'Connor describes the alchemy of her field research in further detail:

> When the process is judged to have gone far enough the spent plants are taken out, first allowing water to drip back into the container. Lime is then added, usually measured in a gourd. The water becomes light blue and is then whisked to incorporate oxygen until it becomes frothy. This makes the indigo pigment, known as indigotin, become insoluble. After being left to settle for a while the water is scooped out of the barrel and the watery paste containing the indigotin is drained into a basket lined with a piece of cloth. The dark blue paste is kept damp and used to make the dyeing vat . . . The insoluble paste will only become soluble [again] under reduced conditions, that is, in water from which oxygen has been removed.
>
> O'Connor 1994a: 69–72

Historic accounts of the preparation and development of indigo are strewn with bodily references, and expressions relating to 'feeding' the 'body' of the vat are commonplace in technical handbooks. In her translation of the memoirs of an eighteenth-century French dyer, Cardon records the dyer's assertion that he must judge from experience whether the vat needs 'feeding with lime and indigo or whether it is fully engorged' (Cardon 2016: 52). Balfour-Paul describes the fermentation of the dye bath – that is, converting the indigo into a soluble, reduced, deoxygenated state with microbes – as 'nourishing the dye vat with organic substances', and she likens the bacterial breakdown of indigo matter to 'leavening bread' (1998: 119). Furthermore, in communities where the fermentation of indigo is accelerated by the addition of human urine as a mild alkali (Balfour-Paul 1998: 124), the process becomes a very bodily procedure in the literal sense.

In 1813, the chemist Edward Bancroft recorded a graphic description of the fermentation of indigo. He wrote:

> In a few hours . . . according to circumstance, a slight motion begins to pervade the body of the liquor in the vat. The bulk increases considerably with some additional heat. Air bubbles are generated, some of which remain on the surface. These gradually collect into patches of froth. A thin violet, or copper-coloured pellicle . . . makes its appearance between the patches, soon the thin film that forms the covering . . . begins to be deeply tinged with fine blue. The liquor has from the beginning, been acquiring a green colour, and now it will appear . . . a bright yellowish green . . . a proof that the base of the colour is now perfectly dissolved in the watery menstruum . . . [and] the fermentation continues [with] much intestine motion through the vat.
>
> Bancroft 1813: 174–5

The life and death of an indigo vat

In Morioka, Iwate Prefecture, Japan, traditional indigo dyeing is practised by artist Ryuta Sasaki. Sasaki is not a successor to a workshop that has been in existence for generations. Rather, he has arrived at his method as a result of research. Shigeru Yoshida, director of the film *Blue Comes from Indigo, Yet it is Bluer than Indigo Itself*, a Fantasia Yoihi Project, 2024, introduces Sasaki's practice as a commitment to 'reproducing the indigo of the past . . . in pursuit of the ideal blue'. Sasaki is particular about the materials he uses, and does not use any motorized equipment such as mixers. Instead, he employs a heavy pestle to 'crush the *sukumo* into small pieces, and then stirs it with his feet for dozens of hours, in the traditional way' (Yoshida 2023).

In the film *Blue Comes from Indigo, Yet it is Bluer than Indigo Itself* (2024), Sasaki describes the process of making his dye vat as 'building the indigo'. He uses just three key ingredients – indigo, water and lye. Sasaki uses *sukumo* supplied by Kitamura from the dried and fermented leaves of the *Polygonaceae* family of Japanese indigo. The process of *sukumo* production is approximately 500 or 600 years old, and the substance forms the base ingredient of all traditional Japanese indigo dyeing. Inspecting a sample of Kitamura's *sukumo* matter in his hands, Sasaki exclaims, 'This is amazing. This is *sukumo* of amazing quality.

There are all kinds of bacteria sleeping in here. I can obtain that beautiful indigo blue colour. It feels mysterious and magical.'

The *sukumo* in Sasaki's studio is mixed with local spring water fetched from the collection point established by the Seiryusui Spring Association. Sasaki transports his buckets and containers of pure water on dollies, wheeling it to his studio where the water is mixed with lye – ash extracted from the burning of oak trees in the region – and a paste is formed. Sasaki explains:

> Japanese people used to burn wood in their everyday lives, so ash was a very familiar substance and they had so many uses for it. But nowadays, to get ash, we have to burn wood just for that purpose. The ash used for indigo dyeing must come from the burning of hardwood. Luckily, a lot of charcoal is produced in Iwate Prefecture. They use a type of oak that produces ash very suited to indigo dyeing. I can get pure and fresh ash, so I'm blessed with a very good environment.
>
> *Blue Comes from Indigo, Yet it is Bluer than Indigo Itself*, 2024

Sasaki is shown collecting his ash from the timber yard. The woodman in Kunohe, Iwate Prefecture, tells Sasaki, 'It's been one and a half years. You can use it right away' and on inspecting the ash in his studio, Sasaki exclaims, 'This is nice ash. Lightly pink.' A solution of water and ash – lye, known as *aku* – is mixed with the *sukumo* to form the paste. Micro-organisms trigger the fermentation process, and the liquid turns blue. After two days, further lye water is added daily to allow the micro-organisms to generate and thrive, and wheat bran, known as *fusima*, is provided as a source of nutrients to encourage the fermentation. Finally, after two weeks of preparation, the indigo vat is 'born' (*Blue Comes from Indigo, Yet it is Bluer than Indigo Itself*, 2024).

Sasaki has developed a close, and particularly personal relationship with indigo, and the poignancy of his emotional attachment is expressed in the evocative description of his work:

> I sincerely love indigo. I love it to death . . . I just can't help but love it. For some reason, I tend to liken indigo dye to a woman. Not from the point of view of a man, but that of a child. I don't say I take colour from indigo. It's more accurate to say that, as a dye, the indigo gives me some of its colour. As I realized that, and adopted a sense of reverence for indigo, I've almost unconsciously begun to refer to indigo as 'Grandma'.
>
> *Blue Comes from Indigo, Yet it is Bluer than Indigo Itself*, 2024

And, as the vat gives up its life and eventually dies, indigo substance becomes a metaphor for being. Yoshida explains: 'Normally, the life span of indigo is three to six months, but in [Sasaki's] workshop, there is a four-year-old indigo that defies common sense' (Yoshida 2023). Shown pointing at one of the four vats in his studio, Sasaki says:

> I hadn't done anything to that dye for a while . . . it just got better on its own. With fermentation you never know. This one's been going on for four years. It's like a 'great-grandma indigo' . . . I don't know what happened. It started to give colour on its own . . . Older, grandma colours have clarity and a greenish hue . . . a colour [that] young dyes can't produce. The result is an exquisite light blue. It's very precious. Just like humans, its condition changes day by day. When it doesn't feel well, it won't give a nice colour, if at all. When it's in a nice and healthy condition it gives a beautiful colour. And when it doesn't feel good, it gives a blackish colour. When it's tired, it just won't give any colour. It goes to show that indigo is alive. It's truly alive.
>
> *Blue Comes from Indigo, Yet it is Bluer than Indigo Itself*, 2024

On a new day as the shutters go up on Sasaki's studio, he says:

> Today's the last time. It's in a worse condition than yesterday. Yep, today's the last time. Usually, when it's not feeling well, I let it rest for a while. But this is the end, so I want to check its colour. I'll give it a little try, but, no, this is it. [She's alive though . . .]

Standing before the vat he bows in respect.

> Thank you so much . . .
> [She had a good run].

The water is emptied and the slurry is lifted from the base of the vat, then bagged up to drain and left to dry into a thick paste. The vat is rinsed out and the residue is washed away. The film's commentary tells us, 'The indigo fulfilled its role as dye, and now waits to return to the earth' (*Blue Comes from Indigo, Yet it is Bluer than Indigo Itself*, 2024).

Towards the close of the documentary, Sasaki reflects on his dedication to the traditional practices of indigo dyeing:

> My sense of self isn't important in all of this. I don't need to search for who I am. I sometimes wonder why I go through all that trouble. What am I doing? Looking at myself, I feel what I do is obsolete. But I sometimes wonder if that's really the case. Maybe what I do is actually at the forefront. I thought I was behind the times. But maybe I'm actually running ahead of the pack.
>
> *Blue Comes from Indigo, Yet it is Bluer than Indigo Itself*, 2024

Indigo as pigment

Balfour-Paul notes that, while indigo needs to be chemically converted by the vat process, the extracted dried pigment can also be finely ground and mixed with a medium to make ink or paint. She explains that some base ingredients, such as egg white, enhance luminosity, while the Mayas of Central America and the Greco-Roman civilizations of classical antiquity combined it with white clay for the painting of frescoes (Balfour-Paul 1998: 207–8). Balfour-Paul records that fixing indigo with an oil-based medium resulted in a waterproof paint. In the nineteenth century 'woad paint' was developed in Germany as a wood preservative in the decorating of houses and churches, while a century later it was used to resurface the sky-blue ceiling of Shakespeare's reconstructed Globe Theatre in London. Balfour-Paul recounts that the Yoruba people of Nigeria mix the remains of their indigo vats into plaster to decorate the walls of their houses, applying it particularly around doors and windows in the belief that the blue tint deters insects (Balfour-Paul 1998: 211–12).

According to Balfour-Paul, staining leather, parchment and paper was first developed by the ancient Egyptians, and the application of indigo in the dyeing, sizing and burnishing of leather was commonplace in medieval times. She describes how the leather was 'dressed with alum and eggs before dyeing it with indigo, normally ground up with whitening, and blended with honey or vinegar, wine or lye, and gum arabic' (1998: 212). On paper, indigo had the benefit of strengthening the fibres, and protecting them from the infestation of insects. In addition, Balfour-Paul explains that, traditionally, indigo dyeing has been combined with other colouring agents to produce purples, greens and blacks, often glazed with rice powder and gypsum and burnished with agate to produce lustrous surfaces (1998: 214). In the same way that indigo is applied to a textile through its fibres as well as in the piece, paper was dyed as pulp or in the sheet. According to Balfour-Paul, paper and cloth come together in the production of blue rag paper, traditionally recycled from sailors' uniforms, but currently reprocessed from remnants of denim (1998: 215).

Indigo: burnishing and polishing

Indigo pigment is a biproduct of its dye. Between each immersion in the lime green-coloured vat, a blue colour develops on the cloth through a process of oxidation; repeated periods in air giving rise to increasing depths of tone. At the same time, an accumulation of indigo on the surface of the cloth results in a residue that dries to a powdery coating – a purple pigment with the potential to be glazed, burnished and polished to a lustrous finish. From this point in the chapter, the exploration is concerned with the properties of indigo that develop after dyeing – a process referred to by Balfour-Paul as 'finishing' indigo dyed cloth (1998: 133).

The polishing of indigo is practised in many world cultures but the process varies from one region to another, and researchers have found it difficult to capture the exact details of its ingredients and technique. O'Connor made three visits to China between 1991 and 1994 to investigate the heavily pigmented surfaces of indigo polished to a lustrous sheen – the glossy outer skin of indigo dyed cloth. In Guizhou Province, O'Connor recorded how indigo cloth is combined with extraordinary substances, immersed in pigs' blood after dyeing, and then in albumen extracted from eggs. Once dried, the impregnated cloth was struck with a wooden hammer – flat on one side and curved on the other – 'a sort of bat' with which the cloth was beaten 'for two or three days' and 'perhaps repeated with more egg white' (O'Connor 1994b). In the village of Judong in the Chinese region of Dong, O'Connor witnessed women washing cloth at the edge of the river. 'They dye and wash it three times', then beat it with a 'very heavy' mallet – 'the end is rounded and a bit blue'. Earlier, O'Connor recorded a similar process in Gaozhen village. She saw 'beating on a heavy stone about eight inches high' and 'red stuff again – hands had red on them' (O'Connor 1993). O'Connor's diaries log a significant event witnessed in Guizhou Province, southwest China. Her notation appears as poetry, capturing the power of performance embedded within it:

> SHINE – for pale blue
> polish
> stone cloth beater
> *shi nan* 'see saw'
> the man stands with one foot on each side
> and rocks the stone from side to side
> the stone 'press' toppled to
> one side when the man first
> got on & he then righted it –
> to stop this happening – he held
> on to the rafters so that he could lower
> himself safely.
> the STONE has line indentations

> where it is pressed against cloth
> the underside
> that bears the cloth has horizontal incised lines
> in the smooth surface.
> to the right
> a wooden roller
> uses shiny wooden roll to calendar
> get lines out
> how long for?
> – if to start does 10 minutes
> of rocking.
>
> O'Connor 1993

In the costumes of the Miao people that O'Connor witnessed, the results of this coating and beating give rise to highly polished surfaces of iridescent colour from purples through to bronzes and blacks. The cloth is crisp, stiffened by its multiple layers of substance. Indigo pigment, albumen, gums and starches become deeply embedded between the woven threads, constraining the drape and forcing the cloth into a rigid structural form of fine pleats. Miao cloth and its transformation from dense pigment to luminous surface demonstrates indigo as the stuff of colour.

Colour as stuff

In 2012, speaking in the *Anish Kapoor: Ann Lewis AO Contemporary Visual Arts International Address* in Melbourne, artist Anish Kapoor gave an evocative description of pigment as 'colour as stuff' to discuss his pigmented works – a description that is borrowed here to express the properties of indigo – a metaphor for the power of pigment to evoke a sensory experience. The physical changes to its material, and its effect on the re-forming of its fabric, are the exemplification of Kapoor's 'colour as stuff' in the context of indigo, thus forming a basis for new meaning in textiles.

In conversation with curator Nicholas Baume to mark the exhibition *Anish Kapoor: Past, Present, Future* at the Institute of Contemporary Art in Boston (2008), Kapoor discussed his early works in pigment. He noted:

> What is important about these works is not that they are made of pigment. The curious thing is that they *appear* to be made out of pigment . . . It [is] said that what you see is what you get. What you see is not what you get! . . . For me the

> illusory is more poetically truthful than the 'real' . . . I believe an object is read through its skin.
>
> Kapoor in conversation with Baume 2008: 40

Baume observes that Kapoor became interested in how an object's surface, 'or what appears to be its surface', determines the way it is read, regardless of, or even despite, the way it is put together. In Baume's view, 'For Kapoor the pigment forms a skin; a term that suggests a more organic presence than "surface". Skin changes, it breathes, it is seamless and elastic' (2008: 19). Referencing the *Anish Kapoor* (2009) retrospective exhibition at the Royal Academy, cultural theorist Homi K. Bhabha discusses Kapoor's early pigment pieces, and asks, '[I]s the skin of the work an organic outgrowth of the form or is it merely a fragile surface dusting of powder? Is the object a remnant or a rendering?' (Bhabha 2009b: 31). Baume describes Kapoor's pigment works as 'floating in the most peculiar way', describing them as 'uncertain objects' with an ever-ambiguous physical state (Baume 2008: 16):

> There are several kinds of uncertainty about the pigment pieces. First is the question of whether or not they are made entirely of pigment . . . then the dematerialising retinal blur of pure colour that makes edges ambiguous and details fugitive. Forms appear to emerge as if from below the floor or beyond the wall, implying that what we see may only be part of something much larger.
>
> Baume 2008: 16

Kapoor explained to Baume, 'Over the years [I] have evolved certain languages . . . the void language, the pigment language' (2008: 40). In 2012, at the *Anish Kapoor: Ann Lewis AO Contemporary Visual Arts International Address*, Kapoor described his making of *Void* (1990). He started with an empty bowl shape 'as if to empty out all content'. When he painted it blue, he began to realize that 'it wasn't about emptiness, it was about filling up . . . I was making a space full of darkness' (2012). Bhabha declares that 'If you think that you have seen "emptiness" as that hole at the heart of the material's mass . . . then think again' (2009a: 172). 'Kapoor's making of emptiness . . . never fails to register a lateral movement, a transitional tremor' (Bhabha 2009a: 171).

In his work *Materials against Materiality*, Ingold challenges us to reconsider material culture as if we were moles, burrowing through material from the inside out:

> Your world would consist of corridors and chambers rather than artefacts and monuments . . . Anthropologically trained moles . . . would doubtless insist that the materiality of the world is not culturally constructed but culturally excavated

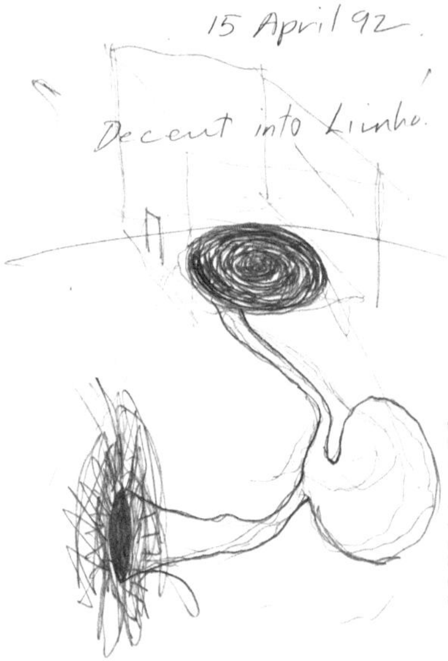

Figure 3.2 Anish Kapoor (1992). *Artist's sketchbook.*

> . . . in the sense that the forms of things are hollowed out from within rather than impressed from without . . . [as] inward facing surfaces . . . as material absence . . . as externally bounded volumes of empty space.
>
> Ingold 2007: 6

In 2014 Kapoor captured the news headlines with Vantablack, a newly developed material – the darkest manmade surface at that time. The blogpost *Super-black coatings* (Brezinski 2016) reports Vantablack as 'absorbing virtually all incident light . . . It reflects so little light that it is often described as the closest thing to a black hole we'll ever see.' With the potential for optical effects, Vantablack is so black that when it is applied to a three-dimensional object it becomes extremely difficult to discern any surface features, and three-dimensional objects appear to become two-dimensional. The super-black material has been developed primarily for scientific and military use. With the potential to hide stealth aircraft, it can block all light from entering telescopes, enabling them to see the faintest of stars in the universe. Kapoor's fascination for the blackest of blacks has resulted in his collaboration with Surrey Nano Systems and his exclusive rights to Vantablack in art. In 2016 in the *Guardian*, Kapoor reflected on our perception of black and its performative power: 'Perhaps the darkest black is the black we carry within ourselves. It's not the night where you switch the lights off – it's the night where you close your eyes' (Kapoor in Delaney 2016). In parallel,

O'Connor's observations of the Miao people – their pigmented surfaces and the performative processes they employ – resonate in the density of Kapoor's black voids – his 'nothingness' (Kapoor in Delaney 2016).

Mud: the passage of time

Mud containing iron is intrinsic to the colouring of textiles. In resist-dyed textiles colour on cloth is transformed by the pigmented surfaces of the earth – organic matter and mud. Together, these substances consist of the residues of plant material and extracts of minerals – clay, iron and tannin – that make up the dyes, pigments and resistant materials inherent in the traditions and practices of resist dyeing.

Traditionally used as a mordant, mud is a 'biting' agent that captures tannins extracted from plant matter and fixes them into the fibres as fast colour. In many world cultures, iron and tannins have been used in conjunction with indigo, resulting in colour combinations of blue tones alongside hues of earthy greys, browns and blacks. Examples are evident across the field of resist-dyed textiles, from Indian chintz to Indonesian batik. In eighteenth-century European calico printing, the range of iron and tannin colours was commonly referred to as 'drab'. In the history of textiles, mud is an intrinsic element of colour on cloth.

In India, mud is also applied as a resist agent, obstructing the penetration of dye into cloth (Ronald 2007). In Rajasthan, *dabu* mud paste consists of well sieved and soaked local black earth combined with tree gum. It is mixed with discarded wheat grains that have been spoilt by insects, but retained and finely ground for this particular purpose. The paste is printed onto cloth from deeply cut wood blocks. To prevent smudging, the wet relief is dusted with fine sawdust or powdered cow dung that results in a series of encrusted gem-like protrusions on the surface of the cloth. Once the mud paste dries, it protects the printed areas from the absorption of indigo dye, revealing jewel-like white marks as the mud is washed away. With the addition of alum in the mix, *dabu* becomes both a resist and a mordant, enabling madder dye to develop into bright pinks, reds and oranges. In contrast, with the addition of iron in Indian block-printed textiles, madder is 'saddened' to a palette of greys, browns, purples and blacks (Bühler 1983). *Dabu* is a product of the earth. As a biproduct of indigenous plant material, *dabu* forms part of an ecological process in Indian textiles.

In Africa, *bogolan* cloth is the result of a complex layering of hand-painted iron and tannins. In the Bambara region of Mali, the local ponds provide mud

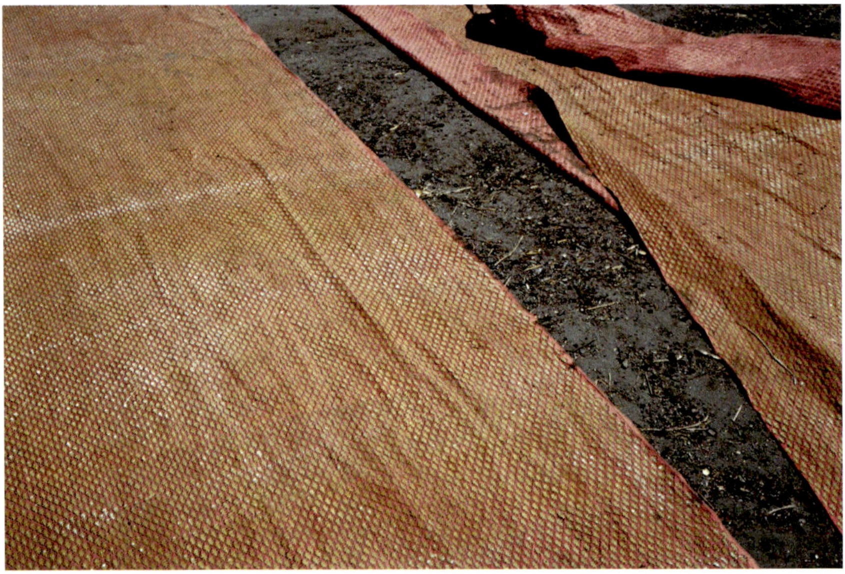

Figure 3.3 Block printed *dabu* mud resist on cloth prior to indigo dyeing, Rajasthan, India, 1989. Photo: Linda Brassington.

rich in iron. This substance is collected and left to ferment for a year. The surrounding leaves and branches of the African birch are pounded and soaked to a brown tea, and cloth, previously steeped in this solution, is painted with the muddy paste. When the cloth dries, the combination of mud on tannin produces a soft grey. As the dyer vigorously rubs the dried surface of the cloth, the powdered mud falls to the ground, back into the earth, and the residual dust is washed away. The process of soaking, painting and drying is repeated a second and third time, each new layer turning the painted areas a darker tone. The remaining yellow parts are painted with a natural bleaching agent – a thick, brown concoction of ground peanuts, caustic soda and millet bran – and the cloth is subsequently placed in the sun. After one week, the bleaching paste is washed away to reveal bright white areas against a dark patterned ground. At this point, the yellow tannins have served their purpose – converting the iron oxide mud into iron tannate, and into the deep, dark dye characteristic of African textile traditions (Toerin 2003). *Bogolan* is the product of the land, a unique relationship between the sun, the earth and the vegetation they yield.

On the island of Amami Oshima, in the archipelago of southern Japan, the climate is tropical and subject to frequent typhoons. Dense layers of iron-rich mud have accumulated over one and a half million years, slowly rising to the surface in swamps scattered across the island. The process of *dorozome* mud dyeing has been practised for more than a millennium. Tannins are extracted from the indigenous *sharinbai* tree, and releasing its tannin content is a sustainable process. The tree's wood chip is boiled for sixteen hours, the remaining mulch is used for firewood and the resulting wood ash is recycled in the practice of indigo dyeing. Then, the hot water extract is left for a week. Spots of bacteria accumulate on its surface, slowly developing into a thick skin of downy mould. Combined with lime, the dye solution produces a 'madder'-coloured pink on fibre, thread, cloth or garment. In the BBC Four programme *Handmade in Japan: Mud Dyeing the Silk* (2017), after mordanting his material in the tannin solution, *dorozome* dyer Yukihito Kanai can be seen taking his stuffs to the swamp. He steps into the mud pit, wading into the paddy up to his knees. He uses his feet to churn up the mud at the bottom of the pool, rotating his boots to create a swirl of grey clay. Pushing his cloth down and under the surface of the glossy viscous sludge, he watches as the textile turns to tones of deep grey-black, twisting it to assess the colour as he wrings out excess clay from its threads, working 'in tune with nature' (*Handmade in Japan: Mud Dyeing the Silk*, 2017). Later in 2018, in the online video *Traditional Amami Mud Dye*, at a place where mountain water meets the stream, Kanai moves his cloth rapidly through the fast flow, washing and rinsing out the mud from the fibres until the water runs clear. Repeated mordanting and dyeing, from as many as eighty to a hundred times, slowly and successively deepens the colour from soft greys to intense blacks. The resultant colour of *dorozome* dyeing is a subtle and variable balance between the solution of tannin extract and the iron-rich content of mud.

Minerals and matter as 'place'

The philosopher and art critic John Ruskin described the surface of the earth as a 'veil of strange intermediate being', a place that 'breathes, but has no voice; moves, but cannot leave its appointed place; passes through life without consciousness' (Ruskin 1860). In visualizing surfaces, anthropologist Tim Ingold has developed a theory that surfaces 'must not only raise being "up" from the depths, but also take knowing "down" from the heights. It must join with the texture of the world, with its materials and processes, answering to them as they to it.' From Ingold's perspective, this is 'a way of thinking and working with stuff' (2017b: 100). In his view:

> The place to start . . . is with the mundane surfaces of everyday life: water, mud, stone . . . paper and cloth . . . [and] with everyday gestures . . . polishing . . . grinding . . . operations that turn rough into smooth, dull into shiny. Many of them involve utensils that remain unsung, scarcely even warranting an entry in any inventory of material culture.
>
> Ingold 2017b: 100

For Ingold, surfaces are important not in themselves but 'for what they potentially open up, and for what they disclose', and he redefines Ruskin's 'veil', stressing that 'The veil covers, but does not cover up. It is not a disguise but a revelation . . . far from hiding the depths *behind* the surface, it allows us to feel the depths *in* the surface' (Ingold 2017b: 104). In this respect, this chapter considers the potential of dyes and minerals, the stuff of colour, as an expression of time and place – exploring the feelings that they can evoke and the meanings that they reveal from beneath their surface.

Indian artist Manish Nai says of his work, 'Blue isn't a colour to me; it is a feeling' (Nai in Mahtani 2018). Nai's 'stuff' is the constricted and distorted material of his personal history in urban Mumbai, the city beneath his feet. His close connection to Mumbai's busy streets informs his work. Nai's father had a wholesale business supplying jute, and Nai spent years watching him mould the material. Since 2017 Nai has returned to the indigo-dyed jute cloth of his family – a material central to life in India. Nai engages large teams of workers. They twist and squeeze copious piles of indigo fabric over wooden sticks, and push it into tall, oblong box constructions. After saturating them with glue, the material solidifies into elongated rectangular forms of compressed cloth. Capitalizing on a culture of mass production by manual labour, Nai's posts of recycled material are produced in multiples, shown stacked or propped up against the wall. At first, they appear as orderly rows of indigo structures, then seem to be falling into partial disarray – a reflection of an overcrowded urban sprawl in semi chaos. The convoluted mass of Nai's work expresses the depth of his material and the density of his home city in a constant state of movement and change. In the *Architectural Digest* of India, Nikhita Mahtani (2018) describes Nai's installations as 'the intricacies of Mumbai – its contrasts, contradictions and controlled chaos, realised on a canvas of recycled material'. In Mahtani's view, Nai's use of recycled jute envelops his exhibitions with a part of his home town, even when they are in far flung places across the world. Mahtani challenges our perceptions of Nai's work: 'Where a viewer might see a grand sculpture, Nai sees every twist and turn in the jute fabric used to make it' (2018). His experimental *Untitled* series (2018–19) uses indigo-dyed jute on tall wooden frames – majestic in scale, yet using everyday materials. Nai says of his work, 'Bombay [sic] is chaos . . . So my intention is to take that chaos and create something that looks inward.' Mahtani observes:

> The dichotomy between the compression of the materials and the grandness of their scale is a direct representation of the contrast Nai attempts to show on a daily basis – that of the hustle and bustle of a major city with the simple, quiet moments that make the space home . . . Most Indian art is loud; it demands to be seen, boasting luscious bright hues and grand canvases. Nai's [work] is still seen, but more so because it hits the part of human consciousness most people fail to portray on the outside: their innermost feelings, who they are once the noise fades away.
>
> Mahtani 2018

In his installations, Nai's use of the depth of tone and texture of indigo is an unveiling of life in the metropolis.

In contrast, Chiyoko Tanaka's work is an intimate engagement with cloth and pigment in the rural surrounds of Kyoto. Tanaka's handmade small-scale works reveal her system of long intense polishing, the physical action of her process. Tanaka dyes reels of wrapped threads in indigo to achieve gradations of deep blues, and weaves the yarn into cloth. After removing her cloth from the loom, she extracts minerals from the earth outside – mud, clay, ochre, charcoal – and forces them into her woven fabric with bricks and stones. In *Grinded Fabric – Ocher* (1991), Tanaka compresses minerals into the structure of her cloth until the indigo-dyed threads break to reveal the depth of material underneath its encrusted surface.

The presentation of Tanaka's work emphasizes its substance, often spread out on the floor in direct contact with the ground, appearing to submerge into the skin of the earth beneath. O'Connor's account of polishing indigo pigment in a remote corner of China is reflected in the intense and deeply immersive experience of Tanaka's 'grinding'. Her bodily exertion echoes the rhythm and repetition of her making, the force of her creative gestures equating with the burnishing of indigo-dyed cloth. For Tanaka, the significance of place is central to her work. In November 2019, at GalleryGallery in Kyoto, Tanaka explained that, while the Garden of Eden is not a concept to which she can relate, William Blake's vision 'To see a World in a Grain of Sand' in his *Auguries of Innocence* (1863) is a spirit that greatly resonates with her. For Tanaka 'outside is best', whether in the splashes of opaque white calcium ground from gathered shells, or in the spots of mud scattered over her weaving. When asked whether a visit to Stonehenge in 2001 had been of particular significance in her work, Tanaka replied, 'All places are important, but their significance is my place in their space' (Kyoto: GalleryGallery, 5 November 2019).

Takeo Uchiyama, a specialist in contemporary Japanese art, summarizes Tanaka's articulation of cloth as 'an accumulation of time' (Millar, Uchiyama and Tanaka 2002: 4). As Tanaka captures time through performative process with materials, tracing the texture of the ground into her cloth, so artist Richard Long captures time as a bodily engagement with the land, making marks as he walks through the landscape, kicking the dust along his pathway and imprinting the passage of time in the matter beneath his feet. His medium is not textiles but, like Tanaka, his material is the earth and its surface – mud. In his work *Dusty to Muddy to Windy* (1993), Long walked 191 miles 'from Bristol to Truro' over five days – encountering 'high tide on the River Parrett', walking along 'dusty lanes up onto Dartmoor' and passing through 'muddy lanes' to 'china clay tips' in 'a following wind', leading him to a place where 'the road [was] flooded' (Long 1993). As Tanaka collects charcoal and ochre to grind, Long gathers clay to paint. Extracting sediment from the rivers of his native Bristol, he mixes mud to different consistencies – as a paste to print, or a liquid to paint. Long impresses the muddy palm of his hand onto surfaces in a sequence of gestures, or uses it as a tool with which to fling, splash or stroke a series of directional marks across the gallery – over the floor and across the walls in rhythmical geometric composition (*Frieze Masters 2013: Richard Long*, 2013).

In 2008 Richard Long stood on the banks of the River Avon with writer and critic Robert Butler: 'I am using the Avon as material. The Avon provides the mud' (Long cited in Butler 2008). Writing in the journal *1843* (2008),

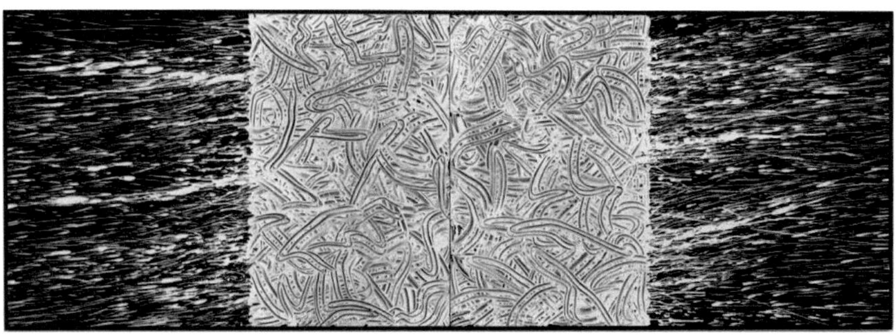

Figure 3.4 Richard Long (2014). *Untitled*. China Clay on plywood. Diptych. © Richard Long. All Rights Reserved, DACS 2025.

Butler noted that 'the raw material for much of [Long's] art lies immediately below, around and in front of us'. Long explains that tidal mud is made by the gravitational pull of the moon over millions of years. 'The River Avon mud is unbelievably strong and viscous. It has all the natural binding qualities, like cave paintings.' Long has used mud from other places, but Avon mud is his 'favourite mud because it's my local mud'. Even its molecular structure is different from the mud in a country field. Long loves its 'squidginess', and he takes this mud all over the world. According to Long, the River Avon has the second highest differential between high tide and low tide in the world, and he finds the spring tides compelling: 'We have a river and every six hours it has no water in it', at which point the river reveals its mud banks, creeks and inlets (Long cited in Butler 2008). In 2018 in the *Guardian*, Jonathan Jones described Long's painted works in clay on linen at the Lisson Gallery: 'His

Figure 3.5 Detail of ochre rubbed over hand-woven ramie and indigo-dyed warp in Chiyoko Tanaka's *'Grinded Fabric – Ocher'* (1991). Photo: Linda Brassington. Photographed and reproduced with the permission of Lesley Millar.

muddy swirls mass in an engrossing swarm. The flow and life of the river seems caught in this whirlpool of mud' (2018).

In Ingold's view (2019a), the ground on which we stand is not a supporting platform but a 'zone of formative and transformative processes set in train through the interplay of wind, water and stone, within a field of cosmic forces such as those responsible for the tides'. Ingold refers to these as 'materials in motion . . . a world in movement, in flux and becoming'. Seen from Ingold's perspective (2017c: 4), the solidity of the ground begins to 'heave and swell'. It is the place where 'we swim in an ocean of materials' (Ingold 2007: 7). These transformations mark the passage of time and place. While Long celebrates the land, revealing the earth's resources accumulated over millions of years, according to Uchiyama, Tanaka's work 'unveils the essence of reality, too often concealed by worldly illusions'. Her grinding of minerals 'symbolises the passage of time'. Furthermore, as areas of her cloth are left threadbare, and the warp previously concealed behind the weft is exposed, Tanaka's work might also be seen as capturing the future (Millar, Uchiyama and Tanaka 2002). Tanaka's textiles are an expression of Ingold's vision of surfaces as a veil through which to discover the earth — a veil that 'far from hiding the depths *behind* the surface . . . allows us to feel the depths *in* the surface' (Ingold 2017b: 104). Together, these haptic experiences reveal the performative power of pigment.

Encrusted

At the *Higeta Aizome Kobo* in Mashiko, pigment is the stuff of indigo, the residue of dye that collects on every surface with which it comes into contact – cloth, earth, the rim of a vat, the sides of a vessel – and in the splashes that fall on the surrounding walls and floor. Indigo is matter that accumulates. It creeps into every corner of the dyeroom, filling cracks and crevices to form a crust of iridescent crystallized purple over time. In the work *Encrusted* (Brassington 2016) these qualities are simulated on cloth. A textured medium of pumice and acrylic on the closely-woven structure of narrow linen cloth forms a mass of granulated relief on the surface. After fifteen successive immersions in a fresh and strong indigo vat, alive with a breathing purple bloom on its pulsating surface, a deep and intense colour permeates through the grains of pumice into darkness. This work is the antithesis of subtle diffusions of dye. Rather, indigo exudes through the cloth. It breaches the cloth's defences, seeking out its vulnerabilities, leaching into its fibres and infiltrating its substance before leaving an encrusted purple residue on the surface.

Figure 3.6 Linda Brassington (2016). *Encrusted* (detail). Indigo, pumice and acrylic on Japanese linen cloth. Photo: Linda Brassington.

4

Process as Performance: Repetition, Ritual, Gesture

Building on the stuff of colour and the metaphorical meanings that material substance can generate, this chapter reflects on the *actions* of resist-dyed textiles, their qualities and effects. The chapter continues the exploration of anthropologist Tim Ingold's definition of making as 'the intervention between fields of force and flows of material' (2009a), asking what are the 'fields of force and flows of material' in resist dyeing, and how can they be seen? Here, the performance of indigo and resist dyeing is explored through three principal themes – as bodily action (the physical engagement in the making of work), as material transformation (in the interventions and interrupted flow of material) and as physical or emotional response (in the actions and reactions provoked by the work).

Contemporary artists' practices, and reviews of their work by other scholars, are at the centre of this chapter. The works are introduced as spaces where new interpretations of indigo and resist dyeing can be explored, where parallels can be drawn between innovative practices in resist-dyed textiles and original concepts in other disciplines. The chapter begins with a comparative study of two of the foremost contemporary indigo artists in Japan – Shihoko Fukumoto and Chiyoko Tanaka – and associates the work of the late Hiroyuki Shindo with British artist Anya Gallaccio's engagement with indigo. It questions the fields of force in the making of their work, distinguishing their bodily actions as controlled or spontaneous gesture, or as conscious detachment – active and passive.

Indigo dyeing is interpreted as a physical intervention in the 'flow of material' (Ingold 2009a), as a transformation of colour on cloth. The chapter questions how this phenomenon corresponds with other material transmutations – crystallizing minerals in the work of Roger Hiorns' *Seizure* (2008), liquefying and solidifying metal in the work of Matthew Barney's *Water Castings* (2015), filtrating fluids in the work of Anya Gallaccio's *All the rest is silence* (1999/2018). These works are discussed to illustrate the parallel practices of contemporary artists who exploit chemical reaction and physical response in the form of sculpture, installation and performance.

Throughout the chapter, the indigo vat is identified as a vessel – the site for the transformation of colour on cloth. The writings of William Morris demonstrate the challenges of physical engagement with the indigo vat, and cultural theorist Homi K. Bhabha defines the studio itself as 'a subliming vessel for alchemy and exploration' – the extended space as the site of material experimentation. Through *In conversation: Matthew Barney and Homi K. Bhabha* (2015), Bhabha introduces the idea that 'viewing the work becomes part of the process of making it' and, from that point, the performativity of indigo and resist dyeing is explored through its physical and emotional resonance. The chapter associates the transformative power of Richard Wilson's oil-flooded installation *20:50* (1987/2018) and Anish Kapoor's voids with the perception of space echoed in the dye vats of Púchov in Slovakia and Mashiko in Japan. Ultimately, this chapter asserts indigo's capacity to generate somatic and psychological effect and demonstrates the power of the dye vat to challenge visual perception.

Resist dyeing as bodily action

Performativity as bodily action is fundamental to resist dyeing. It is demonstrated in the pressure exerted on the cloth through folding or pleating, and in the thread used to stitch through its layers or to bind over its surface. Acts of resist dyeing reflect the manipulations of the thread and cloth that create them. Scholars Yoshiko Wada, Mary Kellogg Rice and Jane Barton describe *Shibori* as derived from the verb *shiboru*, 'to wring, to squeeze, to press' (1985: 7), emphasizing the actions performed on cloth to resist the indigo dye. These acts of resistance are 'imprinted memory' (1985: 54), demonstrated by the amount of pressure exerted in securing them in the dye, and revealed as traces held within the cloth. From Ingold's perspective, when practitioners make things, they 'bind their own pathways or lines of becoming into the texture of material flows' (2009b: 91). He describes the formation of fossils as 'pleats and creases [that] fold in and on each other, matter holding itself together' (Ingold 2019a). His poem 'Fold' (2017a: 72) signifies folding and creasing as a performative process, seeming to echo the bodily engagement with cloth, and encapsulating resist dyeing in his verse. While the verb 'fold' is used singularly, it is also used to indicate what is hidden between the layers. It is both neutral and loaded with meaning:

> Fold
> Where side by side is back to back or face to face
> What secrets lie between the sheets of bed or newspaper
> Where words like bodies touch and kiss in unseen intimacy?
> To read, the pages must be opened up, and words that once
> had felt
> Each other's pulse
> Must stand apart as though they'd never known each other,
> Divided by a crease.
>
> Ingold 2017a: 72

The imprinted memories of resist dyeing, therefore, can be perceived as demonstrations of intimacy and gesture – wrapping as containment, binding as restriction – a bodily engagement with cloth.

Curator Mary Jane Jacob (1993) acknowledges Fukumoto as one of Japan's leading exponents of *shibori* and indigo dyeing, learning from tradition while reinterpreting the craft as contemporary art. Wada, Kellogg Rice and Barton (1985: 7) describe how the variables and influences of resist dyeing 'conspire' to remove some of the human control in the process, but Jacob recounts that Fukumoto found the rich, somewhat unpredictable nature of the dye endlessly fascinating, 'intoxicating' (Jacob 1993: 9). Fukumoto works with the natural fields of force in resist dyeing and with indigo's potential to change and develop on the cloth. She applies her acts of resistance by piercing a single stitch through layers of tightly packed cut cloth, or by compressing a systematic series of tight folds. Fukumoto understands her cloth. Her relationship with it develops through touch. She feels the packing of its threads, the tightness of its weave, the density of its substance. Her senses enable her to anticipate the results of her dyeing, the extent to which dye will penetrate the multiple layers of cloth through a single stitch, crease or fold. Fukumoto's understanding of indigo informs her dyeing – her ability to judge the period of immersion in the vat; the sequence of lowering and raising her cloth in and out of the dyebath; the pulling and releasing of the winch to achieve subtle gradations of colour and tone. Ingold visualizes this relationship as 'a thing and its shadow', elements working in collaboration, 'going along together' (Ingold 2019a). He stresses that 'All are part of each other's stories, not adjoined, but *conjoined*, melding together. In the shadow of material

Figure 4.1 Detail of Shihoko Fukumoto's indigo-dyed cotton sateen over stitch resist (2001). Photo: Linda Brassington. Photographed and reproduced with the permission of Lesley Millar.

things are materials in correspondence with each other, in a gathering formed by compression' (Ingold 2019a).

Fukumoto is recognized for her early resist-dyed works of numerous repeating bright, white dots that accumulate across her vast expanses of dark cloth. Her *Constellation* (1996) is a large-scale composition containing multitudes of radiant lights, appearing as distant stars against the night sky. Fukumoto's works are installations that envelop space. As part of the education programme supporting the exhibition project *Textural Space* (Millar 2001), Fukumoto demonstrated her process in a group workshop. The day was employed in the systematic layering and cutting of cloth strips into deep stacks of compacted squares. The dense piles of cloth were locked together with a single stitch – a single piercing, knotted and tied. At the point when the wet package was subjected to indigo, it was swiftly removed, vigorously rinsed and plunged into a dryer – the vacuum drawing out the excess moisture as it spun, forcing air through the cavity and developing the rapid oxidization of colour at high speed. Fukumoto's results were multiple layers of blue cloth with a central bright, white light – a pinprick on the outside, gradually enlarging to a diffused radiant spot on the inside, the light repeating as it moved down through the layers of cloth to the centre of the pack (*Untitled*, 2001). In Ingold's view, 'A world of flux is a folding and creasing of a sponge like surface' (2019a), and in the context of Fukumoto's resist dyeing, the penetration of indigo through the cloth is determined by her bodily actions and sensory understanding, coupled with the structure of her cloth and thread. Together, they establish the 'fields of force and flows of material' (Ingold 2009a) fundamental to her work.

Art historian Marjan Unger (1992: 28) distinguishes the work of weaver and dyer Chiyoko Tanaka as 'special' among her contemporaries. Tanaka's practice is reclusive. She works in her remote house and studio in the forest, a simple small-scale dwelling integrated into the surrounding nature. It is a sensory space inside and out. Tanaka moves through her workspace barefooted, feeling the treadles of her loom and sensing the surface of the ground underfoot. Tanaka's work is a bodily engagement with material in the literal sense. She develops an intimate relationship with matter, from dye to thread to cloth. After dyeing and weaving, she takes her cloth outside, to a rock sited in the forest outside her studio, a vast flat surface forming a raised platform. Tanaka lays down her cloth onto the surface and kneels before it, smoothing its threads and ensuring that the cloth is in contact with the surface of the rock. She crumbles a heap of dried pigment onto the woven structure – clay, ochre or charcoal – and rubs it with a stone on stone. It is a slow rhythmical process, leaning forward and pushing the stone away from her body before drawing it back again in repetitive motion. Tanaka grinds the minerals through the surface of the cloth, tracing the ground and transferring its texture to the fibres 'to embed, implant, erode or emboss its own surface into my work' (Millar, Uchiyama and Tanaka 2002: 29). Ingold claims:

> . . . a back-and-forth movement of the hand and arm that is neither precisely measured out nor targeted . . . but harbours an intense and intimate 'feel' for the materials of which these objects are formed . . . by hand . . . registers every bump or hollow, every crease or fold, not as a feature set upon the surface . . . but as a variation intrinsic to the surface itself.
>
> Ingold 2017b: 101

Tanaka generates a friction that raises the fibres of her cloth, rubbing the surface to the point where it becomes threadbare. Unger writes that '[a] careful observer sees how the irregularities in that base recur in her work', in the way the cloth has been worn down and compressed, and in the holes that are 'essential' to the work (1992: 28).

Tanaka's textiles are the embodiment of performativity in cloth, and Ingold's anthropological view of the haptic is manifested in her work:

> We can at last return to the gestures and implements . . . with which we began . . . polishing, grinding . . . and so on . . . operations that smooth the surface . . . that bring one surface into contact with another: interfacial not in the sense of crossing a threshold between the exterior and interior of an object . . . but in the sense of establishing a relation *between faces*.
>
> Ingold 2017b: 104–5

As Tanaka lifts her cloth from the rock, and allows the air to flow through its substance, the indigo-dyed reverse is revealed and light passes through the distressed holes from back to front, indigo reappearing through the density of mineral deposit on the upper side of the cloth. In Tanaka's work the cloth becomes the interface between her physical action and the rock, her body and the environment. As the piece is removed, only the traces of mineral extending beyond the selvedge remain – a symbol of time passing through Tanaka's very existence.

Performativity in Tanaka's work is the result of both bodily action and material transformation. In Unger's view (1992: 28), Tanaka's textiles are 'material out of which another material is created'. The act of rubbing, or 'tracing' – the process of 'grinding' the cloth – translates the texture of the ground into Tanaka's 'canvas'. In *Six Squares, Indigo Blue* (1994) Tanaka has built her surfaces and then broken them down to reveal their hidden material qualities as illustrated on the Browngrotta Arts webpage *Chiyoko Tanaka*. As Lesley Millar, former Director of the International Textile Research Centre, observes, 'as we see the erosion, so we understand the continuity contained within the material, the process, the place and time' (Millar et al. 2002: 12). Millar also notes that '[t]he importance of allowing ourselves time to look [at the surface revealed], cannot be too highly stressed' (Millar et al. 2002: 14). Tanaka's *Six Squares, Indigo Blue* is a dense surface where minerals have become bonded together within the woven structure beneath, a new substance where warp, weft, dye and mineral have melded together to become one. Ingold stresses that:

> Haptic vision . . . abides with surfaces, and dwells in them. Its interest is less in the conformation of the surfaces than it is in their *texture*. And this texture tells not of the form of things but of their substantive composition.
>
> Ingold 2017b: 102

Although it cannot be touched, *Six Squares, Indigo Blue* evokes a profoundly haptic experience.

Observers of material flow

In comparison with Fukumoto's systematic folding and Tanaka's material change through erosion, in the film *Textile Magicians* (1997) the late Hiroyuki Shindo applies indigo in a single gestural splash. Shindo binds polystyrene globes with strips of indigo-dyed cotton cloth overlaid with hemp yarn, rotating the spheres in his hands and building numerous layers of dense lines over their surface. He rolls each orb of wrapped threads over the surface of the vat and then holds it static, allowing the indigo to be absorbed into the surface, upwards and through the threads from dark indigo to white, forming soft shadows behind the undyed yarn. Shindo is shown carrying his wrappings to the river. He crouches to roll the textile globes in the moving water, allowing the force of the current to redistribute the spheres in the riverscape, tossing and bouncing them on the surface of the water in an animated performance of dyeing.

In complete contrast to these vigorous actions, Shindo devised a slow, and strikingly different, system of resist dyeing using a shallow trough lined with stones. Rather than applying forceful gestures, Shindo created dips and hollows between the stones, smoothing areas of cloth into the depressions and allowing the remaining mounds of stones to form gentle resists against the indigo which he slowly trickled into the exposed cavities beneath. As the dye was absorbed in variable degrees, a surface of graduated light and shadow materialized. On the Browngrotta Arts webpage *Hiroyuki Shindo* (2017), Shindo has described the transformation: 'I watch carefully as the indigo is drawn slowly into the fabric, the resulting gradations of hues of blue – from nearly invisible shadows to areas of nearly black.' Ironically, Shindo told us, 'I have discovered that the white in each work . . . is as great a concern as the dyed portion. If the white is not brilliant enough, or the undyed portion is not of the right proportion, the balance is broken.' For Shindo, the point at which the process stopped was significant: '[W] hite is as important to my work as indigo' (*Hiroyuki Shindo*, 2017).

In *Textile Magicians* (1997), Shindo is seen exploring the flow of indigo in the roundness of his spheres and across the horizontal surface of his cloth. In contrast, British artist Anya Gallaccio exploits gravity – creating *All the rest is silence* (1999) in the vertical. Commissioned by Sadler's Wells, the 1999 wall panel accompanying the work described Gallaccio's process: 'Like the shows in

Figure 4.2 Anya Gallaccio (1999). *All the rest is silence* exhibited at Sadler's Wells. Chappa silk, indigo dye, watercolour paper. © Anya Gallaccio. All Rights Reserved, DACS 2025.

Sadler's Wells' theatre, you are witnessing here a rare performance, a slow motion drama: in this case the slow dyeing [with indigo] of a long bleached curtain' (Visual Arts at Sadler's Wells 1999).

In 2018 the original work was recreated at the John Hansard Gallery, Southampton. According to the gallery webpage Gallaccio's installation 'evolves through a process of transformation', indigo dye 'slowly suffuses' down a 7-metre length of white chappa silk in front of the gallery window overlooking Guildhall Square. The gallery's press release described Gallaccio's concept as W. H. Auden's poem 'The Sea and the Mirror' (1944), a commentary on Shakespeare's *The Tempest*.

> . . . this world of fact we love
> Is insubstantial stuff:
> All the rest is silence
> On the other side of the wall . . .
>
> Auden 1944

This work is an intervention on cloth, but Gallaccio distances herself from the material, stepping back from the process and allowing indigo to gravitate, slowly dripping through the fibres over days, weeks and months. The temporality of Gallaccio's work is emphasized on the gallery webpage: 'Indigo conveys the magic and mystery of performance.' On 23 November 2018 the gallery announced on Twitter: 'The first trickles of indigo mark the opening' (John Hansard Gallery 2018). Posts read: 'Wonderful to see . . . beguiling . . . with indigo seeping out into the nocturnal landscape of Southampton's Guildhall Square' (Moreton 2018), and by January 2019 another proclaimed: 'Quite a dramatic change since my last visit' (Martin 2019). The following account describes the experience:

> *On a wintry afternoon in February 2019, looking from the gallery window to the surrounding cityscape, there is a heavy fall of snow. The sky and the ground are converging in a vast blanket of whiteness across the city. Slowly, very slowly, trickles of indigo run down the folds of white cloth against the backdrop of snow outside. Drips escape, scattering over the pristine platform beneath. As the gentle flows of dye meet the air, they change, their colour fixed in time, their momentary journey locked into the fibres of the cloth. Over time, deep blue has permeated Gallaccio's silk, saturating it with graduating tones, dark and then flowing through the cloth to light below.*

At the John Hansard Gallery, indigo was presented as an event, as a momentary performance of dye. Over a period of ten weeks, indigo gradually descended through the wall of cloth until the exhibition drew to a close.

Ingold describes practitioners as 'not so much imposing form on matter as bringing together diverse materials and combining or redirecting their flow in the anticipation of what might emerge' (2009b: 94). Both Shindo and Gallaccio stand back from their work, exploiting the material's propensity to flow, and follow its natural course in a slow performance of change. Although their works are carefully rehearsed manipulations – staged events from the outset where each artist anticipates the outcome – Shindo and Gallaccio become passive observers of material 'flow', allowing the natural 'fields of force' (Ingold 2009a) to determine their final outcome.

Performativity as material transformation

British artist Roger Hiorns transforms material through chemical process. Hiorns' description of the making of *Seizure* in 2008 – his crystallization of copper sulphate as an intervention of an empty space in London – draws extraordinary parallels with the chemical and physical processes of indigo. In 'Interview: Roger Hiorns on Seizure' for *Artangel* (2008), he discusses how the innate qualities of matter, organic substances with their own physical properties, have the potential to 'distance the work from the control of the artist's hand'. Here, the chapter examines this principle in relation to indigo, and questions how its material qualities form the making of work.

In the online video *Roger Hiorns, Seizure 2008/2013 at Yorkshire Sculpture Park* (2013), Hiorns describes the transmutation of material into form, the process in harnessing, not an object, but a state of change as art:

> I was thinking about taking what I make in the studio and wanting to scale it up, using material which didn't have any sign of my own hand, setting up a system and letting that system dictate or contain its own aesthetic decisions . . . a system that is just doing what it does if you give it the right environment . . . There seemed to be a universal truth to the idea that you would just let nature, under the right circumstances, do what it needs to do . . . Once you started the process up, once you had the right proportions, like temperatures, it seemed an unlimited process . . . and then it came to be apparent that it wasn't an *object* to be crystallized, but something stranger.
>
> *Roger Hiorns, Seizure 2008/2013 at Yorkshire Sculpture Park*, 2013

In Hiorns' making of *Seizure*, suppliers of copper sulphate shipped containers of the blue powder from Italy to a disused council bedsit in Southwark, London. The space was sealed and filled with hot copper sulphate solution and the chemical left to cool in its dissolved state, 'taking its temperature daily'. It lost a degree a day, 'we thought it would take longer', and after approximately one month, when the cooling had reached an appropriate level 'we simply sucked the liquid back out again'. It became apparent that the crystallization had occurred, and 'on a scale that we had not anticipated' (*Roger Hiorns, Seizure 2008/2013 at Yorkshire Sculpture Park*, 2013).

In 2008, in a review of Hiorns' *Seizure* for *Icon* magazine, William Wiles heralded the result as 'astonishing': 'The initial effect is captivating – every surface is covered with thousands of sparkling crystals, all of them a remarkable deep blue. It's like entering a cave encrusted with precious gems, or walking into a huge geode . . . Splendid and beautiful' (Wiles 2008). Wiles goes on to describe the visiting of caves as a practice invented in the eighteenth century by 'thrill seekers looking for a newly identified sensation, the sublime'. Wiles associates the sensation with the uncanny, a place where 'the sublime rests with ambivalence . . . danger made delightful' (Wiles 2008):

> At first *Seizure* provokes joy; presently, however, that changes . . . This sense of perversion stems from the air of catastrophe that clings to *Seizure*. Something has happened inside this flat, and however wonderful the results may seem, they carry an implicit threat: you don't want it happening again, you don't want it happening to you . . . Danger is in the air . . . The crystals are reaching out towards you; they have grown, like plants, but they are not plants. If the implacable, uncaring disinterest of nature can be mistaken for malice, wait until you see how hostile minerals can seem . . . The invasion of the flat is infection of the inhabitant. Contagion is in the air. This patch of . . . modernity . . . has been seized.
>
> Wiles 2008

Wiles continues: '*Seizure* strays into the mystical. It's a space that has genuine spiritual depth' (2008).

In 2013 the work was extracted from its original site and transported to the Yorkshire Sculpture Park in Wakefield where it is relocated. Now, entombed in a new vessel for its longer-term preservation, a protective shell prevents the crystals from degrading over time from the wet and from the heat of the sun. The continuity of *Seizure* is ensured for a minimum of ten years. In *Roger Hiorns,*

Figures 4.3–4.4 Roger Hiorns (2008). *Seizure*. Copper sulphate crystals. An Artangel/Jerwood Commission, Harper Road, London. © Roger Hiorns. All Rights Reserved, DACS 2025. Image courtesy Corvi-Mora, London. Photo: Marcus Leith.

Seizure 2008/2013 at Yorkshire Sculpture Park (2013), Hiorns describes *Seizure* as being 'in semi storage' in a building designed 'to allow this limbo to continue'. Ingold reminds us that 'makers have to work in a world that does not stand still . . . and with materials that have properties of their own and [that] are not necessarily predisposed to fall into the shape required of them, let alone to stay in them indefinitely' (Ingold 2009b: 93).

In 2009, when *Seizure* was shortlisted for the Turner Prize, its significance was not the nomination itself, but the fact that Artangel commissioned the work not knowing the result. Co-Directors Michael Morris and James Lingwood committed to the project before the results of that crystallized space could be seen. 'Artangel has always gone where others fear to tread. We collaborate with artists who defy boundaries to give form to extraordinary ideas . . . It's often unlike anything you have experienced before' (Artangel 2020). *Seizure* (Hiorns 2008) signifies a commitment to the many variables and unpredictability of material transformation, embracing risk and uncertainty as a fundamental element of all making. *Seizure* is the essence of Ingold's 'fields of flow' (2009a), the epitome of process as performance.

The subliming vessel

In contrast to the slow transformation of material in the work of Shindo, Gallaccio and Hiorns, American sculptor Matthew Barney's bronze casts assert a different power – a force, a single violent gesture. From here, this chapter develops a comparative study of Fukumoto and Barney through their contrasting actions in the transformation of materials. Barney's material is fluid. Unlike Fukumoto's indigo, his molten metal triggers a sudden and extreme reaction when subjected to water. Thierry Somers (2015) explains that Barney's molten bronze is poured into a pit of bentonite clay silt. As water is poured into the vessel, the liquid metal 'violently displaces the water and finds its way into the crevices between the pieces of clay', solidifying to a random sculptural form.

In 2015 Barney exhibited *Matthew Barney: River of Fundament* at the Museum of Contemporary Art in Los Angeles (MOCA). The exhibition consisted of a work in film alongside a series of sculptures – his *Water Castings*. In *In conversation: Matthew Barney and Homi K. Bhabha* (2015) at MOCA, Bhabha describes Barney's *Water Castings* as 'a rare insight into the making of work'. Having worked in collaboration with the Walla Walla Foundry in Washington over a period of seven years, Barney describes each casting as 'a very specific choreography' (2015). The results are explosions as form, reactions as performance. In their conversation, Bhabha refers to the water castings as 'transformative moments . . . moments of art as ritual'. In his view, the foundry allowed Barney to think about the studio as 'a subliming vessel for alchemy and exploration'. Barney defines this as a place where materials are 'allowed to change on their own terms, and to make certain demands', a place where 'the synergy that forms [them] is greater than the sum of its parts' (*In conversation: Matthew Barney and Homi K. Bhabha*, 2015).

In Barney's practice, Bhabha redefines the studio itself as the 'subliming vessel' – the extended space as the place for the transmutation of material – but Fukumoto's workspace is the quintessential example of this metaphorical description. While Barney's foundry is the setting for explosions in the transformation of metal, Fukumoto's vessels are the site for the transformation of colour on cloth. Millar explains that Fukumoto has devised approaches particular to her practice. Tools and techniques are her own invention, 'allowing her to respond intuitively to the possibilities inherent in the dyeing process' (2001: 49). Each work develops in relation to another, in parallel, in multiples, as series. Millar writes, 'Constantly experimenting and testing each new piece lays down the sources of inspiration for the next' (2001: 49). Fukumoto's pattern of approach is demonstrated in her workspace. Dye vats line one wall with pulley mechanisms above them. Millar records that Fukumoto's indigo dyeing can extend over a period of four to five days, immersing the cloth tens of times to achieve the desired depth of indigo. She shifts the work between one strength of vat and another, spraying waves of water over its surface to create subtle gradations of tone over the cloth (Millar 2001: 49–50).

In comparison, Somers' description of Barney's approach is equally representative of Fukumoto's practice:

> Barney is part of a group of artists who are makers and doers and their work has got a physical quality . . . the creative process is visible and traceable in the final result. It is 'Generative Art' [using] systems of chemistry, biology and mechanics where 'happy accidents' can occur. The final result cannot be predicted as the work unfolds and makes us realise that most of the beautiful things in art occur when you let them happen.
>
> Somers 2015

Through her understanding of living material, Fukumoto coaxes the development of her colour on cloth, exploiting the capacity for change through biological fermentation and chemical reaction.

Jacob explains the term *kogei* – the Japanese concept of craft as art. *Kogei* is about dedication, 'no matter how laborious and time consuming it might be to master a technique or execute a work' (Jacob 1993: 7). Having trained as a painter and worked in steel, glass and plastic in her early career (Daiwa Anglo-Japanese Foundation 2004), Fukumoto found indigo through marrying into a family of indigo dyers and by studying the process of natural indigo production from plants. In Jacob's view, Fukumoto was not interested in 'making change for the sake of being revolutionary but in conveying the idea of "Japanese" in a new way, through self-expression' (Jacob 1993: 9). Fukumoto's route to self-expression was through tradition but without the restrictions of convention.

It is significant that, in an interview with Randy Kennedy of the *New York Times* (Kennedy 2015), Barney describes how, at the time of the *Water Castings*, he felt he had 'worked himself into a conceptual corner'. His sculpture was closely tied to the narrative in his films and he 'couldn't see a compelling way to continue' (Kennedy 2015). When structuring his film *River of Fundament* around Norman Mailer's 1983 novel *Ancient Evenings*, Barney noted that the book was set at a time when the Egyptians were making great advances in metalworking, casting and the development of alloys (Kennedy 2015). Kennedy recounts this as an important influence in shifting the direction of Barney's practice:

> This gave him a way in to begin thinking about materials . . . to think about a project in which sculpture would drive filmmaking . . . Mr Barney said, 'I've always had an interest in colliding the natural and artificial, but I think what makes this work different from what I've done before is that the natural is foregrounded.'
>
> Kennedy 2015

In their contrasting ways, Fukumoto and Barney exploit the performativity of material transformation. It is the gateway to their creative production, the vehicle for their vision.

Performativity as physical and emotional experience

Indigo is an evocative substance. From this point, performativity is considered as physical and emotional experience. As textile historian Gillian Naylor notes, William Morris's description of the dyehouse in nineteenth-century England (1876a) presents a daunting image: 'The vat you must know, is a formidable looking thing, 9ft deep and about 6ft square: and is sunk into the earth right up to the top' (Morris in Naylor 1988: 130). In her comprehensive history of *Blueprint in Slovakia* (2014: 106), Olga Danglová calculates the depth of the indigo vats as 1.5 to 4 metres and containing up to 3,000 litres of indigo. Her description of the vat as 'look[ing] like a deep basin' recalls memories of the indigo vats at the workshop of Stanislav Trnka in Púchov:

> *Trnka peers down into his three concrete vats in the floor of the dyehouse. Deep pools of dye send overpowering smells of fermenting indigo into the enclosed room. The purple scum on the surface of the vat is pulled to one side. Slowly, gently, the winch is turned above the indigo and the cloth enters the dark watery abyss, disappearing and rising again until the drips subside. In the darkness of the dyeroom, the indigo vats exude an overpowering smell of ammonia from their depths. A continuous spiral of cotton cloth has been wound around a metal star frame. The crank is turned and, as the chains shift, a loud series of clanking, clunking and clattering emanates across the room, while the giant metal form slowly submerges beneath the ominously dark depths of the dye liquor. The splashes stop, silence pervades, until, after one hour, the heavy dripping form is raised above the surface and the sea green cloth turns a shimmering blue. Light pierces through the wooden slats of the dyehouse, catching the wet cloth and glistening on the surface of the vat.*

Thirty years later, the same indigo vats remained buried in the empty workspace:

> *Now the dyeroom lies motionless and dry, drained of its liquid content. But the smell of dye and chemicals still lingers, permeating its ethereal presence through the blackness, and resonating a deep sense of melancholy and abandonment.*

Trnka recounted an 'unpleasant incident' in the 1990s. During a school excursion to his workshop 'an inquisitive schoolboy fell into the vat and almost drowned' (Trnka in Danglová 2014: 106). Indigo triggers an emotional energy and resonance that can present physical danger – the lure of its liquid mass and the awe of its inner darkness. Entering Hiorns' *Seizure* and stepping over the threshold from broad daylight through to a very dark space – the point where the adjustment to darkness takes place – is significant. In 2013, audiences entering *Seizure* at the Yorkshire Sculpture Park were given guidance: 'Overshoes must be placed over existing footwear . . . take care when entering . . . there is a step . . . walk slowly and carefully . . . the floor is uneven . . . mind your head . . . surfaces are sharp . . . do not touch . . . gloves are available to offer protection should you fall.'

This is the point where, in Bhabha's view, 'the material and the non-material tangentially touch'; a phenomenon that Bhabha illustrates in the works of British artist Anish Kapoor, where the fragile balance of doubt and uncertainty are experienced in the deep spaces of his surfaces (Bhabha 2009a: 172–3):

> A sudden disappearance of surface in a deep, dark hole literally cuts the ground from under our feet; the body loses its direction and density; the eye hovers, horizonless, homeless. Each step we take towards the work places us on moving ground; each time our gaze is suspended between the frame, a lip, of fine luminosity surviving just above the shadow line, and then all view is lost; vertigo.
>
> Bhabha 2009a: 173

In 2018 British artist Richard Wilson's seminal work *20:50* (1987/2018) showed in *Space Shifters* at the Hayward Gallery, London. The exhibition was described as bringing together 'sculptures and installations that explore perception and space . . . [and] engage with the way that the act of looking involves the whole body'. From here, the exploration presents an analysis of *20:50* (1987/2018) in relation to the resonance of Japanese modern and contemporary resist-dyed cloth, underpinned by the overarching perception of the indigo vat as a void.

The heightened sense of smell and the unsettling perception of space in Trnka's dyeroom is manifested in Wilson's seminal work, *20:50* (1987/2018), and the emotional power of Wilson's installation is echoed in the indigo vat as a

Figure 4.5 Installation view of Richard Wilson's *20:50* (1987) at *Space Shifters*, Hayward Gallery, 2018. Photo: Mark Blower. Reproduced with the permission of Richard Wilson.

sensory space. Both are fluid in content and immersive in materials that saturate the mind. The *20:50* gallery space was filled to waist height with recycled engine oil. A walkway led from a single entrance, taking the viewer into the space surrounded by oil on all sides. Visitors to *Space Shifters* were issued with instructions: 'Do not touch the artwork, doing so will cause permanent damage to clothing and other personal belongings . . . Visitors with long hair are advised to tie it back . . . young children should be held . . . do not hold cameras over the edge in case of being dropped' (Hayward Gallery 2018). Writing in the *Guardian* in 2003, Wilson described an earlier moment when a woman visited *20:50* at County Hall, London, wearing a white Burberry coat and carrying a baby wrapped in a white shawl:

> I looked at her gleaming white clothes, and for a moment I was really tempted to warn her. I hesitated, and she went into the gallery, and clearly didn't quite know what was going on. She leaned out over the oil, trying to work it out, and her hair plopped down and she immediately threw it back, all over the white Burberry coat. It was the most unbelievable mess. I went in and took the baby, and gave her tissues and turps and apologised.
>
> Wilson 2003

Wilson described the inception of *20:50* as a fascination with 'a Tardis-like space, where the internal volume is greater than its physical boundaries', the ways that space can be changed 'to alter your perception, to knock your view of the world off kilter'. The oil became part of the piece because he 'knew it had a highly reflective surface. There had been a drum of the stuff sitting in my old studio, without a lid . . . and I always loved the way that . . . there was this void, this perfect reflection' (Wilson 2003), a space Wilson described as 'a perfect black hole' (*Richard Wilson at Space Shifters: In the Gallery*, 2018). The following account describes the experience of *20:50* at *Space Shifters*:

> *Waiting in the queue of visitors for an hour, the heavy smell of sump oil permeates through the space, between the concrete galleries, around the curve of the stairwell and along the human snake of waiting people. As the queue moves slowly and the work reveals itself from behind the white dividing wall, a vast mirrored surface of oil reflects the light to a silver finish, duplicating and reversing the exact image of the internal structure – glass, concrete, metal – but also reflecting, in perfect detail, the stains, impurities and imperfections accumulating on the weathered areas of the building outside. They contrast with the still, clinical image inside, the semi-toxic substance revealing a beauty in its installation while reflecting an imperfect reality on its surface.*

20:50 was exceptionally compelling but more than visual trickery. It generated a powerful fear of falling, a compulsion to look down into its infinity pool.

> *There is an inability to judge the height, distance or depth of its void, to determine its horizontal plane, standing below the waterline while seeing above it. Only the door breaks the distance, grounding the perception of space until the reflection prevails and presents a deep and overwhelming abyss ahead. Treading the walkway, stepping deeper into the space, a gradual incline triggers an irresistible urge to stop dead, to touch the sides to stabilize, until prompted by the recurring memory of the guidance not to touch, not to grip the edge to rebalance. Retreating, there is the urge to turn again, to catch a final glimpse of the unsettling presence, the absence of gravity, the emptiness.*

In *20:50* only the red wheel on the tap below anchored the liquid in its space and re-established the viewer in a grounded reality.

> *Inside the space, there is an overpowering feeling of isolation and detachment in the stillness. Walking out and re-entering the queue, a strange sense of security is contained within the waiting populous, a reconnection with humanity and an extraordinary feeling of reassurance.*

Wilson's fluid material reverberated in the indigo vats in Púchov, where the vessel had presented a menacing void. And the indigo embedded in the earth at

Figure 4.6 Cloth suspended from bamboo stretchers in an indigo vat at *Higeta Aizome Kobo*, Mashiko, Japan, 2016. Photo: Mayumi Tsukimura.

the *Higeta Aizome Kobo* in Mashiko, Japan, transports the viewer into an uncanny existence.

> *On arrival at the Higeta studio, I meet a dark, potent atmosphere, full of fermenting indigo and thick with woodsmoke under a thatched roof. It is immediately intoxicating. The sun catches the lip of the vats and the purple gloss shines on the surface of the water. The deep ceramic pots are slowly heated by burning woodchip. Their liquid content breathes a cluster of bubbles that accumulate into an effervescing purple mound. As the temperature increases, they multiply into a dense mass of fermenting froth on the surface of the liquor. Each indigo vat is buried beneath compacted earth, forming a curved edge with an ingrained halo around its perimeter. Each perfect circle contains indigo and its constituent parts, determining the condition of the vat, its activity and results; its life. The vat defies perception. As I peer down into its liquid void, there is no sense of depth or space. Even though I know the physical height of the vat, my senses tell me something different – I cannot see beyond its surface. As I work with it, and my hands move the purple bloom to one side, the indigo appears dark, thick and opaque. Only when the cloth is gently raised to the surface does the indigo vat reveal its watery mass, and light, transparency and life reinvigorate the colour on cloth, the sharp green dye slowly intensifying in air from deep to deeper tones. Each piece is swept up from the depths of the stainless steel waterbath. Bending, stooping, crouching; I slowly lower the cloth until it is submerged into the dense cavity of colour. Stillness. Waiting, watching, counting time.*

Martin Gayford of the *Spectator* observes that '[a] reflection on still water was perhaps the first picture that *Homo sapiens* ever encountered' (2018). He also notes:

> Reflections can do two intriguing things. They can make what's really there disappear and, perhaps simultaneously, create a new setting that doesn't actually exist at all. Richard Wilson's celebrated piece . . . does both . . . which adds an undertone of existential threat and perfectly rational anxiety . . . in its injection of a hint of menace.
>
> Gayford 2018

In 2015, Artcourt Gallery in Osaka exhibited *Fukumoto Shihoko: Japan Blue*. It marked the occasion of Fukumoto's forty years of practice in pursuit of her 'ideal spatial consciousness'. Fukumoto's indigo 'ties time-space and beauty that science cannot grasp' (*Fukumoto Shihoko: Japan Blue*, 2015). In London in 2010, at No.15 Royal Arcade, Fukumoto showed indigo-dyed textiles at Galerie Besson. Rupert Faulkner, formerly Senior Curator, Japan, in the Asian Department of the Victoria and Albert Museum, London writes:

> The quality of light in [the] gallery is very special. It streams in from the west through double height windows overlooking Albemarle Street, while to the east another expanse of glass is set against the sugar plum confection of the Royal Arcade. Simplistic a statement though this is, I am always amazed at and excited by the way in which works of art are born out of what seem like quite primitive . . . environments. It must have something to do with childhood memories of being taken on a tour of a car factory in 1960s Japan, or of being shown around the engine room of the ship on which I once travelled from Yokohama to Southampton with my mother and sister . . . It was thoughts of this kind that went through my mind when, upstairs in the room above the [indigo] dyeing workshop, Fukumoto showed me piece after piece of her exquisitely dyed textiles. There is a transformative power about everything she makes, whether a scarf . . . or a massive sequence of hangings that totally redefines an interior or even . . . a whole building.
>
> Faulkner 2010

It was her move into installation at the 1990 *International Biennial of Tapestry* in Lausanne where Fukumoto made use of light and transparency – creating gentle tones of indigo over vast swathes of gauze, brushing the fibres to blend diffusions of colour over its surface. In Jacob's view, 'the spatial presence of these works ties them to Western installation art, while their sense of movement connects them to Japanese textiles like banners and *kimono* that become fully activated with use' (Jacob 1993: 11).

It is Fukumoto's use of space that sets indigo on a new trajectory in form and sensibility. Ushiro Katsuhiko's photograph of *Tea Room – Mist* (1990) shows Fukumoto's indigo cube at Nogata Small and Medium Enterprises University in 1996. The work is constructed from dyed linen, suspended from a lacquered metal frame. It appears as a softly draped structure mounted on a blue platform edged with bamboo. The following account is an interpretation of the work and the feelings it generated from the observation of *Tea Room – Mist* (1990) in a photograph:

> *Inside, a man leans forward over a tea bowl. Two women watch, seated, dressed demurely in kimono – one in light blue, the other in deep indigo with a scattering of intense white spots across her garment. The figures are seated on mats of thick white handmade paper, the deckled edges apparent against the dark platform. The group is enveloped by gradations of indigo from light to dark. The work is a poetic interpretation of space, with a mysterious unearthliness in its performativity. The figures seem submerged behind a diffusion of shaded linen, appearing to sit beneath*

a sharp line of white light. Their faces and upper bodies are veiled by the deep sea of darker tones that surround the cube, until the sides seem to rise up and meet the sunlight filtering down through the cloth circle.

Jacob describes Fukumoto's installation as a work that captures 'the transcendental quality of tea' and 'the spiritual essence of indigo' (Jacob 1993: 11). From Jacob's perspective, 'Entering the womb like space, one is enveloped by the blue of the universe and infused with the spirit of Japanese culture' (Jacob 1993: 11). But, in its partial concealment, this seemingly beautiful depiction as observed in the photograph might also be interpreted as an unsettling encounter.

Indigo is performative in its material and process, and in its powerful and emotive qualities. The works of Fukumoto are concerned with space either through the environment they inhabit or in the world they reflect. It is through the medium of indigo that their forms are created, casting shadows while enhancing light, shaping surfaces, and expressing depth and space. Each work creates a unique combination of materials and elements to create an extraordinary environment, a new world that generates a sublime experience, challenging our perception of space, and dramatic in its emotional resonance and audience response.

Referencing the works of Anish Kapoor, Bhabha determines that transition allows 'time and space to develop its own affects – anxiety, unease, restlessness – so that viewing the work becomes part of the process of making [it]' (2009a: 172). This sensation resonated in the dye vats at the Higeta workspace where the ground, the rim of the vat, the reflections on the indigo became forms of transportation to an imagined world:

The deep, dark, intensely pigmented vats are encrusted with residues of salt and mineral deposits, crystallized chemicals around their rim, a hard pitted layer of bronze. I perceive the earth around them as a primeval landscape; a strange and alien terrain in which lakes of indigo have become trapped and embedded as pools of colour. Deep cracks and crevices around the Mashiko vats bear the marks of time, age and continuous activity.

In characterizing material as an organic transitional medium, Bhabha captures the essence of performative process evident in bodily action, material transformation and emotional response – each one performative in its 'fields of force and flows of material' (Ingold 2009a). Bhabha stresses this characteristic as a sublime *something*. This *something* can be interpreted as a metaphorical expression for the resonance of indigo and resist dyeing:

> Material is like living tissue ... its relational powers reside in an on-going temporal process. The process of 'making' does not stop with the manufacture of the object [but with the ambition] to make work that is more than its moment and other than its maker . . . true making often finds itself in . . . the in-between temporality, that *something* – the strange sublimity of technique – that locates the object between the static and the dynamic, in a transitional state.
>
> Bhabha 2009a: 174

Capturing these qualities as images was central to practice at the *Higeta Aizome Kobo* in Mashiko in 2016, and they formed a key element in the resulting exhibition of work at the Mashiko Museum of Ceramic Art in 2017 (*The Exhibition of the Works by Linda Brassington*). They are a visual record of the 'intervention between fields of force and flows of material' (Ingold 2009a), and an expression of a personal encounter with the indigo vat, the 'subliming vessel for alchemy and exploration' in resist dyeing (Bhabha in *In conversation: Matthew Barney and Homi K. Bhabha*, 2015). Together, the collection of photographs is a vehicle through which to contextualize the process of indigo dyeing and a representation of the emotional and psychological experience of the residency:

> *The depths of the multiple indigo dye vats, the smoke rising from the wood fires that heat them, the cracks in the encrusted earth that surround them are all-consuming. The immersive experience they generate, and the compelling sensations they trigger are profound – the bodily engagement with the substance of indigo, the performativity in its material transformation on cloth, the power of the void that permeates from within the dye vat.*

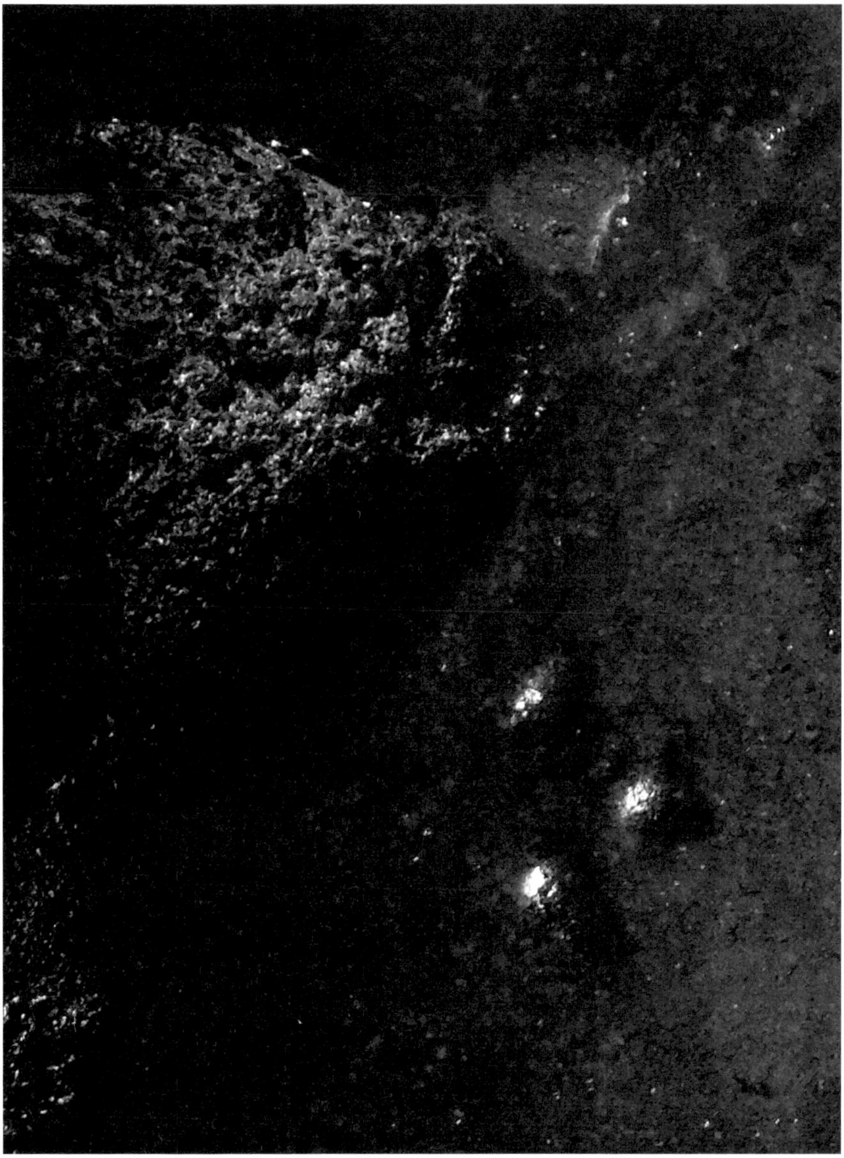

Figure 4.7 Fermenting indigo dye at *Higeta Aizome Kobo* in Mashiko, Japan, 2016. Photo: Linda Brassington.

Figures 4.8–4.9 Work in progress at *Higeta Aizome Kobo* in Mashiko, Japan, 2016. Photo: Linda Brassington.

Figures 4.8–4.9 (*Continued*).

Figure 4.10 Indigo vat at *Higeta Aizome Kobo* in Mashiko, Japan, 2016. Photo: Linda Brassington.

5

Theatres of Making: Space and Place

This chapter is a personal reflection of making in the context of place. It explores indigo and resist dyeing through art residency, working collaboration and field studies in Central Europe and Japan – locations that have been the subject of experiential research since 1985. The research began with a Churchill Fellowship in former Czechoslovakia and developed with subsequent visits in 1986 and 1987 supported by the British Council. Thirty years later, in response to the loss of traditional dyers in Slovakia, the research in Central Europe extended to Austria to study indigo and resist dyeing as a living craft. In 2016, an intensive period of research began in Japan as a result of an invitation from the Mashiko Museum of Ceramic Art to undertake an art residency at the 200-year-old site of indigo dyeing located at the heart of the pottery town. Since then, sites and subjects have been revisited, and the investigation has developed significantly through a growing synergy with contributors to the experience and a repeated personal engagement in their sites of practice. As such, this chapter is both a chronology of research over thirty years and a choreography of craft through workspaces in Central Europe and Japan.

Through an understanding of the geographer Yi-Fu Tuan's ideas of space and place, this chapter reflects on the power of place in resist dyeing. Tuan draws a distinction between space and place: 'Space is abstract. It lacks content; it is broad, open and empty, inviting the imagination to fill it with substance and illusion . . . Place, by contrast, is the past and the present' (1975: 164–5). Consequently, this chapter includes reflections on places no longer active, and others where practices continue while referencing the past – the heritage of their site, or traditions that inform their processes. As the chapter begins, parallels are drawn between the ideas of Tsuyoshi Tane in contemporary architecture and the ephemeral installations of artist Chiharu Shiota, considering how memories might be revealed through searching, surveying and storytelling. The textile works of Reiko Sudo, whose Japanese heritage links directly to her design practice and collaborations with traditional craft, reflect the connection between memory and innovation.

Tuan defines place as 'time made visible, or place as memorial to times past' (1977: 179), and in the context of resist dyeing, space and place are explored in the layers of indigo embedded in its environment, both physically and metaphorically. The chapter examines how sites of practice convey materiality, and how spaces reflect performance and gesture, now and in the past. It questions how empty space can capture a sense of lacunae as a hiatus in time – a point between the past and the present – and how traces of movement are held in the workspace, continuing to evoke the sensory experience of resist dyeing over time. The exploration considers the significance of the workspace as a microcosm of memory in the surface and structure of its building, a place where 'objects anchor time' (Tuan 1977: 187). As such, the chapter reflects upon indigo's depth of space, and examines how a sense of place is evoked in craft heritage sites, described here as 'theatres of making'.

Anthropologist Tim Ingold associates the past with the layers of the earth and their movement beneath our feet. Like the earth, heritage builds up, information is memorized, but slowly forgotten and buried in layers over time. Knowledge and understanding are 'eroded' (like the impact of the weather on earth) but they rise again in a cyclical process of 'opening and closing'. Ingold's analogy is based on the notion that 'things emanate back to the surface'. He perceives heritage as a fluid movement, a flux that is constantly 'rolling, turning and churning' (Ingold, 2019b). Significantly, in his lecture *What on Earth is the Ground?* Ingold presented heritage as a phenomenon that is *in* the earth, not on it. Heritage has 'volume' (Ingold, 2019b). As the chapter unfolds, this analogy is important in bringing different dimensions to the definition of intangible cultural heritage in resist dyeing, opening up new perceptions of the subject.

Memories of place

Speaking at the Daiwa Foundation in London in October 2019, Tane described his architectural design process as 'archaeology for the future' (Tane 2019). As buildings become taller and expand upwards, Tane is 'trying to find something we have lost'. Methods of 'search and research' and 'digging and building' are fundamental to his design philosophy, methods that 'delve into memories of place' to inform architecture of the future:

> We begin our work like archaeologists . . . we start to explore a long distance in time and excavate memories of . . . place. It is a process of surprise and discovery,

> of searching to encounter what we didn't know, what we had forgotten, and what was lost . . . memories . . . deeply embedded beneath the ground . . . from the past to the future, the memory of . . . place slowly becomes architecture.
>
> Tane 2019

Whether digging, mapping or envisioning the ground, all Tane's projects explore the significance of space and place in imagination and creative practice.

A collaboration between Dan Dorrell, Lina Ghotmeh and Tane resulted in *Kofun Stadium* (2012), a design created in response to an international competition in Tokyo's bid for the 2020 Olympics, of which drawings and models are recorded on the *Atelier Tsuyoshi Tane Architects* website. The shortlisted design proposed a manifestation of the ancient burial mounds of Japan. Tane described the concept as the preservation of a mountain in the city combined with the ancient culture of building villages on a hill – settlements in the foothills that mark the boundary between the mountain and the arable flatlands below. The project proposed the planting of a forest to create a micro environment in the city centre of Tokyo, a stadium built by the community for the people. In November 2018 in the *Japan Times*, Sachiko Tamashige commented on Tane's project for Tokyo: '*Kofun Stadium* is so appealing that seeing it causes a slight tinge of regret as to what could have been in store for the capital' (Tamashige 2018). Although *Kofun Stadium* did not materialize, the site of the largest *kofun* site in Japan (*Mozu-Furuichi*) was designated a UNESCO World Heritage Site in 2019.

Tane differentiates between history and memory. According to writer and product designer Shuhei Senda (2018), Tane refers to history as 'selected information that has been arranged by somebody, or many people, that cannot change'. But through his architectural practice, Tane explores the memory of a place left behind in the traces of its history. In his view, by looking into memories of place, architecture can contribute to creating a future: 'I realised that memory does not belong to the past . . . [Rather] memory is a driving force that creates something for the future. That's why we have the knowledge and capacity as human beings to read memories and to connect them with the future' (Tane cited in Senda 2018).

In this respect, Tane and Shiota hold common values of place. Shiota's work is concerned with the concept of absence as presence. In 2017, in the online video *The Theatricality of Chiharu Shiota's Art: Brilliant Ideas Ep.52*, Shiota discusses *Inside – Outside* (2009), an installation of discarded window frames from the empty buildings of East Berlin. The work is a compilation of windows through which residents had viewed the Berlin Wall that had previously divided their city: 'There was nobody, but I still felt the presence of the people who had lived there. The buildings were still breathing.' Similarly to Tane, Shiota wanted the viewers of *Inside – Outside* to experience 'this kind of in-between feeling'. In 2010, on the island of Teshima in the Seto Inland Sea of Japan, Shiota exhibited her artwork *Farther Memory* (*Farther Memory* webpage), a tunnel assembled from 600 windows gathered from abandoned houses on Teshima and the neighbouring island of Naoshima. Ironically, while *Farther Memory* (2010) was originally created as a 'permanent' exhibit on Teshima, Shiota's work is slowly decaying (*Teshima Project*, 2019). As time passes, *Farther Memory* continues to break down, its organic material gradually giving way to the elements before it eventually collapses into the earth. While Shiota's work was created as an expression of memory, its own existence has become vulnerable and its future is in question. *Farther Memory* itself is becoming a 'memory' embedded in the surface of the island.

In 2019, Masataka Hosoo, twelfth generation of the Hosoo family and their long-standing Nishijin textile brand, presented an exhibition of 'skill and beauty that have withstood the test of time'. The exhibition *The Story of Japanese Textiles* (2019) shared tradition among a wider audience 'as a reference for the next generation'. It was staged in their new gallery in the city centre of Kyoto, a space for the portrayal of textiles as 'the history, culture and future of Japan'. In its design and construction, the new Hosoo Gallery has become a receptacle of Kyoto's memory. Its walls are constructed from four local clays extracted from the earth in four different regions of Kyoto Prefecture. Here, the building itself is an expression of Kyoto embedded in the materiality of its structure. In addition, Kyoto's craft heritage is echoed in the bespoke ceramics accompanying the refreshments served inside. The plates and vessels are the products of Hosai Matsubayashi, sixteenth generation of the Asahiyaki family of potters. His limited-edition artefacts are derived from earth in the Uji district of Kyoto, using clay dug up, stored and matured over a period of fifty years. The layers of coloured clays that circumnavigate the perimeter walls encapsulate the Hosoo Gallery, rising up in repeating horizontal bands to the top of the building. Here, earth is an expression of place, the very foundation of Japanese craft.

Figure 5.1 Clays extracted from four regions of Kyoto Prefecture in the construction of Hosoo Gallery, Kyoto, Japan, 2019. Photo: Linda Brassington.

Reiko Sudo

Reiko Sudo, Design Director of NUNO (meaning 'cloth'), collaborates with Japanese craftspeople to preserve traditional textile techniques through contemporary design and production. In 2019 Sudo developed *Jellyfish*, a cloth manipulation celebrating the heat-sensitive, shrinkable and mouldable qualities of synthetic fibres. In 2002, in *Memory on Cloth: Shibori Now*, Yoshiko Wada discussed developments in the heat setting of *shibori*. Sudo's *Jellyfish* builds on these earlier innovations. In 2019, in the online video *Sudo Reiko: Making NUNO* to accompany her exhibition, Sudo explained the process. Initially, small fabric experiments were created – stitching rows and grids on polyester cloth, and baking them in domestic ovens. After much trial and error, she creatively adapted the design for large-scale production through the ingenuity of Japanese silk screen printers. Sudo explored the heat-shrinking qualities of polyvinyl chloride, and combined this material with the heat setting polyester fabric. By screen printing a water-soluble glue that penetrates two layers of cloth, Sudo bonds the polyvinyl material onto polyester organdie. The attached materials are subsequently baked in an industrial oven. The heat causes the polyvinyl to shrink, while the attached polyester cloth is heat set into 'puckering forms'. Once the layers are separated, *Jellyfish* 'squirms with buoyant life'. The result is a simulation of stitched *shibori* in mechanization. Sudo's design innovation demonstrates her understanding of textile heritage and its potential for larger-scale production in contemporary practice.

The influence of heritage in Sudo's work is expressed in her 2019 exhibition *Sudo Reiko: Making NUNO Textiles* at CHAT, the Centre of Heritage, Arts and Textile in Hong Kong. The accompanying online video describes Sudo's *Koi Currents* (2019), a textile installation of ninety carp streamers suspended at the entrance, as 'arranged in a whirling school of modern day *koinobori*'. Traditionally these cylindrical cloth wind tunnels are flown in Japanese gardens to symbolize 'strength, endurance and to herald the prowess of [a] newborn'. Inside the exhibition, Sudo has composed a display of her personal collection of *kimono* fabric swatches 'arranged to reflect the five ancient colours used for dyeing in Japan'. They consist of blue-green, red and yellow from plants, black from earth, and white from water and ash, as noted by Hara (2007: 6). From Sudo's perspective, by tracing their histories and origins, and learning about their weavers and dyers, the swatches continue to inspire her designs. In the online video, Sudo recalls:

> I grew up in a small country town where every spring and autumn we looked forward to the arrival of the travelling salesman with his load of *kimono* fabrics.

> He would present these beautiful textiles one after the next on the *tatami* mats for my grandfather sitting up front, with my mother and aunt looking on from behind. Probably that was when I first dreamed to one day become someone who makes beautiful fabrics . . . my childhood dream . . . behind my becoming a textile designer.
>
> *Sudo Reiko: Making NUNO Textiles*, 2019

In Sudo's dream, heritage and art are in correspondence with each other.

Artistic Director of NUNO, Seiichi Saito, uses technology – film, video, sound – to introduce the stories behind Sudo's textiles. In the online video *CHAT Winter Programme* (2019), Saito explains his objective – to show how the fabricators and people in the factories deal with the processes in reality and on a larger scale: 'I feel like I have to film, otherwise the process [will] disappear. I feel really responsible now, [so] I have to keep doing this.' In Saito's view:

> Technology need not occupy the forefront at all times. A machine, as complete and whole in itself, is a beautiful affair. Such was conveyed to me by the collaborative processes and textile work of *NUNO* and its craftspeople across the nation of Japan. Their old machines were carefully repaired time and again to continue to make beautiful textiles, their sounds churning loudly and ceaselessly through towns in textiles.
>
> *CHAT Winter Programme*, 2019

In the video, Sudo expresses her design philosophy and sets out the story of a *NUNO* textile:

> The most important thing is, even before a textile is made, there will already be many encounters with different people, and these encounters lead to the creation of something new. The textiles we produce are not crafts, although our hands are involved in the production process . . . [but they] highlight the connection between people engaged in the production of these textiles.
>
> *CHAT Winter Programme*, 2019

In search of *ton-byan*

Looking up at her 2019 Hong Kong exhibition installation, Sudo declares: 'Textiles are something that makes us wonder' (*CHAT Winter Programme*, 2019). In 2020 she was filmed in the NHK television documentary *The Lost Textile of Ryukyu*. Recorded amid a global pandemic between periods of lockdown, and at a time of rapid change, the film follows Sudo's journey to Okinawa to explore the island's textile heritage and the continuity of its traditional practices. Sudo wonders, 'What is it that doesn't change?' The impact of the Second World War is evident throughout the film, and Sudo explores Okinawa's motivation to restore textile traditions such as *bingata* – the painting of dye on cloth. The search for renewal amid the devastation of postwar Okinawa is symbolized in the use of spent military cartridges as improvised tools. Still in use after seventy-five years, they serve as nozzles in the application of resist paste onto cloth. In the figurative compositions of *bingata*, colours are held within the boundaries of these resist-pasted lines. In parallel, the dye used in *bingata* is extracted from the *fukugi* tree. Many were destroyed by military bombardment but their seeds were planted in the disrupted earth where they have re-established the roots of their heritage. After 100 years the tree will be mature enough for the extraction of dye. Now, after seventy-five years, Sudo perceives them as a symbol of renewal represented in growth – the 'message to keep looking forward, no matter the hardships'. Sudo traces the pursuit of researchers, weavers and scientists searching for *ton-byan*, 'the lost cloth' of Okinawa. Existing examples of the cloth are unique, extraordinary in their sheen and transparency: 'The first time I touched it, the crisp texture and the lustre of the threads captured my heart . . . No-one weaves this cloth today.' Sudo asks why *ton-byan* did not survive along with all the other textile techniques of Okinawa (*The Lost Textile of Ryukyu*, 2020).

The Lost Textile of Ryukyu (2020) presents Okinawa as located at the crossroads of international trade. According to Basil Hall, a visiting British naval officer in the nineteenth century, the island was perceived as a peaceful utopia. In 1994, weaver Kyoko Shukumine researched 100 textiles from Okinawa in the collection of the Berlin Ethnographical Museum. She describes the enormity of feeling she experienced during her six weeks of intense examination: 'It was like meeting the people behind the textiles . . . Looking at the threads spun so neatly through the fabric, I could visualize the people who worked on it . . . the story behind each piece. It was such an emotional experience.' Shukumine discovered seven examples of *ton-byan* in the collection. They 'shimmered like glass', and yet the documentation that accompanied them gave no record of their material. There are conflicting theories about their fibre content. *Ton-byan* means 'agave' in local Okinawan dialect, while other bast fibres such as banana and ramie have also been investigated. But analysis shows that the *ton-byan* fibre is flat and not plied, and its substance has remained a mystery. Recent scientific investigations suggest that *ton-byan* fibre is related to a nettle native to China. Perhaps the fibre was imported on trade routes between the East China Sea and the Pacific Ocean (*The Lost Textile of Ryukyu*, 2020).

The *Lost Textile of Ryukyu* tells a story of lost heritage, and yet an appreciation for the rarity and unique qualities of *ton-byan* continues to resonate in the human imagination. In the words of computer scientist Mitsukuni Matayoshi, 'If something is "lost" I want to find it.' But from Sudo's perspective, 'There is a mystery [in *ton-byan*] that even today's science and technology cannot solve.' The intangible cultural heritage of Japanese *ton-byan* is held within the secrets of its material substance.

Hranovnica: a hiatus in time

In Slovakia in 1985, the disused indigo workshop of Elemír Montško in Hranovnica stood still. The print room was draped in black cobwebs. Wooden rails were stretched across the ceiling, holding rows of dark, rusted nails where printed cloth would have been hung to dry. A print table for block printing resist paste onto cloth would have accommodated a deep shelf beneath it – a sleeping place where mattresses would have been unrolled for travelling apprentices. They would have worked for short periods as and when the demand required, learning the craft while travelling between one village and another.

Now, the dark and empty room holds a pause in time.

The Hranovnica workshop, once an important centre of production, has become an archive of disused tools and equipment. It is a lifeless repository, a testimony to a lost craft, forty years on from the death of Montško in 1974 and the demise of his practice here in Hranovnica. Indications of past activity are identifiable in the space:

Sunlight makes a brief entrance through the open window in the empty space of Montško's printworks. A grass path outside, still bearing the markings of a boundary now removed, leads to the entrance at the rear of the house.

A series of rooms reveal clues to its former domesticity; the stove in the entrance hall, the kitchen sink, the locked wardrobe, the stencilled walls, the net curtains at the window, the strip of cut paper decorated with Slovak dancers along the edge of the shelves where the finished cloth was once stored. This house was multifunctional; a home for living, a space for meeting visitors – customers who came to select, order and buy their patterned cloth. A warping mill, a spinning wheel, a distaff deposited in the building are reminders of the obligatory flax spinning and weaving that once occupied almost every house in the district.

Tuan (1975: 152) describes place as 'a centre of meaning constructed by experience' but, while places are 'points in a spatial system . . . they are [also] strong visceral feelings'. In his view, 'While it takes time to form an attachment to place, the quality and intensity of experience matters more than simple duration' (1977: 198). The following passages seek to convey notions of emptiness, memory and lacunae experienced during a return visit on a single day in the summer of 2016:

In Hranovnica the workshop, now disused, abandoned, static, holds evidence of past activity, recent change and successive interventions. Montško's workspace, once the site of alchemy, wooden forms and printed image, is motionless. The space is bare apart from a tall twisted paper sack weighted on the floor by its content of dry goods, a single fossilized figure, a distorted petrified form. The walls are empty but for a set of shelves set into the recess, still containing a large glass jar and two further sacks of chemicals; perhaps chalk, gum, copper. They are slumped, the paper dry and crisp, their brittle edges separated into fine layers of fragile fibre.

In the dyehouse a framework of pulleys and supports straddles the width of the room. The star frame around which multiple layers of cloth had once been hooked has dropped to the ground. Above, stained with indigo and wrapped in rusting chains, wooden rollers are imprinted with deep indentations from the incessant rotation of metal links over their surfaces, yielding to the constant pressure and weight of wet cloth. The two dye vats beneath them have collapsed into a single black hole, a deep gaping chasm underground. Bright sunlight stops at the entrance

Figures 5.2–5.3 Details of decay in the workshop of Elemír Montško, Hranovnica, Slovakia, 2016. Photo: Linda Brassington.

Figure 5.4 Interior view of the earth floor at Elemír Montško's workshop, Hranovnica, Slovakia, 2016. Photo: Linda Brassington.

to the washroom, a small dark space with a crumbling stove and washtub. The rusted door to the wood ash below remains open, while the brick and rendered perimeter of the hollows above are disintegrating into the cavity beneath. Fallen plaster, broken floorboards and rotting door frames have begun to break down the structure of the building, revealing its material content as fragmented matter.

In the barn, old latches and temporary locks create apertures through which sunlight is cast onto the stone walls and grey grit floor. The space bears the vestiges of time. These are the first signs of a building succumbing to its surroundings, imploding back into the earth beneath, returning the fabric of the building to physical substance, to pigment, to stuff. Now, it appears as if the dry ground is rising, slowly submerging the abandoned tools, dismantled equipment and stacks of wood stored on its surface. Trails of footprints in the dust lead to dead ends, turning back on themselves and descending into ever decreasing circles; appearing lost and disorientated; ineffectual pathways leading nowhere. And the earth beneath the workshop lies breaking and crumbling to a fragile dustbowl.

Indigo evokes a deep sense of space in its dark heavily-pigmented surfaces. Its material qualities are echoed in Hranovnica and resonate in the contemporary workspaces associated with its processes. The substance of indigo is embedded in

Figure 5.5 Interior view of the pulley in the dye room of Elemír Montško's workshop, Hranovnica, Slovakia, 2016. Photo: Linda Brassington.

Figures 5.6–5.7 Collapsed dye vats in the workshop of Elemír Montško at Hranovnica, Slovakia, 2016. Photo: Linda Brassington.

Figures 5.6–5.7 (*Continued*).

Figures 5.8–5.9 Interior views of the earth floor at Elemír Montško's workshop, Hranovnica, Slovakia, 2016. Photo: Linda Brassington.

Figures 5.8–5.9 (*Continued*).

the very fabric of the buildings it occupies, holding the cultural heritage of generations past and present, and perpetuating the continuous rhythms of making over time. In Tuan's view, 'To know a place is also to know the past' (1975: 164).

Púchov, Slovakia

In 1985 at the Trnka workshop in Púchov, Slovakia, resist-printed cloth was attached to a star frame stretched in a spiral formation. Its selvedge was hooked on alternate pins. The open spaces between the cloth allowed air to move freely through them after indigo dyeing, hastening the process of oxidation and ensuring that the colour developed from greens to an even blue:

> *A rapid beating ensues. Trnka thwacks a stick between the dripping folds, separating the layers of cloth and allowing oxygen to penetrate the fibres. One hour later, after the drips subside, the damp cloth is removed from the frame, reversed, re-stretched and immersed again. The process is repeated and repeated in recurring rhythm; dipping, dripping, beating and turning through as many as eight cycles to achieve a deep blue black.*

In Púchov the print workshop and indigo dyehouse were located at opposite ends of a long woodyard. This space served a particular purpose. Many of Trnka's cloths included two layers of printing, each with two depths of indigo tones. After the initial dyeing, the first layer of paste would continue to preserve areas of white pattern. This needed to remain undisturbed. The cloth could not be ironed to remove the creases, and yet, to avoid unwanted pickup and blemishes forming on the cloth, its surface had to remain flat, ready to receive the next layer of overprinted pattern. A solution to the problem was devised:

> *With Trnka at one end of the courtyard, and his wife at the other, the couple pull the cloth in unison, tugging the ends, yanking out the creases and forcing the cloth into submission.*

After a second layer of printing and overdyeing, Trnka's bare hands agitated the cloth in a solution of diluted sulphuric acid, slowly breaking down the paste, rinsing and dissolving it away to reveal the white and pale blue pattern beneath the dark blue ground. The cloth was taken outside to the river, to the row of trees planted at regular intervals along the bank. Trnka hooked the length of one selvedge onto long nails located at the same height and the cloth was left to dry – a practice repeated daily by three generations of his family before him.

In 2017 a return visit to the disused Trnka workshop revealed the same nails in the trees.

> *The rusted nails still remain in Púchov. Small and insignificant, and almost hidden against the bark, they project from the trees – a lasting testament to past practices until eventually decaying and falling into particles of powdered rust.*

Now, the regeneration of indigo dyeing in Slovakia is the work of Trnka's grandson, Peter, symbolized in the form of a newly constructed dye vat in the ground (*Modrotlač Peter Trnka*, 2014), the potentially transformative space of a subterranean cavity rising up through Ingold's layers of heritage to the surface of the earth (2019b).

Steinberg, Austria

In the fertile agricultural region of Burgenland, Austria, Josef Koó continues his family's practices of hand block printing and resist dyeing. A 200-year-old workshop in Steinberg is the site of *Original Blaudruckerei Koó* (Original Koó Blueprint). Its sense of place in the history of Austrian 'blueprint' is marked by the multiple doors and windows of its buildings, each one layered with paint

Figure 5.10 Helena Trnková and Stanislav Trnka (out of view) pulling out the creases of printed and dyed cloth in Púchov, Slovakia, 1985. Photo: Linda Brassington.

Figure 5.11 Stanislav Trnka drying cloth from nails in trees at the edge of the riverbank in Púchov, Slovakia, 1985. Photo: Linda Brassington.

stained with indigo. The following passage recalls the first steps into the workshop on a cold morning in February 2017:

> *A series of footprints appear on the newly driven snow that has fallen this morning. Their tracks mark the route of multiple pathways from the house to the garden and across to the outhouses. These traces of movement on the ground, and the stains of indigo left behind in the virgin snow, denote a fleeting moment in the transformation of space into place, a point where the relationship between craft and memory is manifested in a pattern of footprints. This is a place where the invisible is made visible, appearing as an ephemeral image of making, before disappearing through the melting ice and back into the ground.*

In Tuan's view:

> The modest work of human erosion . . . the bare earth packed firm by human feet are not planned, but they can be touching. Intimate experiences . . . easily escape our attention . . . It is only in reflection that we recognise their worth. At the time we are not aware of any drama; we do not know that the seeds of lasting sentiment are being planted . . . Humble events can in time build up a strong sentiment for place.
>
> Tuan 1977: 143

In Steinberg, a collection of blue aprons is suspended from wooden hooks at the door to the indigo dyeroom. Entering the space triggers a surge of sensory perceptions:

> *Koó rests his spectacles on his head over a knitted hat. He is dressed in blue from head to foot, wearing a denim jacket over his jeans and overalls to keep out the cold. The three circular concrete vats are buried in the floor. Their rims are caked in multiple layers of dye and chemicals. Splashes and streaks of indigo stain the wall behind, and drip onto the timber slats around them, soaking into the wooden platform across the floor. The perimeter of the space is scattered with paraphernalia essential to the process – a wooden pail, a metal bath, a bucket propped on the edge of the vat, its plastic surface enveloped by a thick purple crust of crystallized indigo. Koó leans forward to turn the crank of the metal spiral frame, pushing the lever with the full weight of his shoulders. He removes his spectacles and places them alongside a mallet, a scattering of hooks and a collection of jam jars containing traces of concentrated indigo liquor. Alongside them, a wooden box contains a pile of circular metal weights that keep the cloth submerged beneath the surface of the dye. The cloth is immersed. The sounds of the rotating metal chains, and the gentle splash of dye liquor as the heavy frame enters the vat, emanate through the room. Then, as the chains drag the frame above the vat, excess dye drips down into pools of indigo on the surface, and the colour on cloth slowly changes from acid green, to turquoise, to blue.*

Figure 5.12 Exterior of *Blaudruckerei Koó* at Steinberg, Austria, after snowfall, 2017. Photo: Linda Brassington.

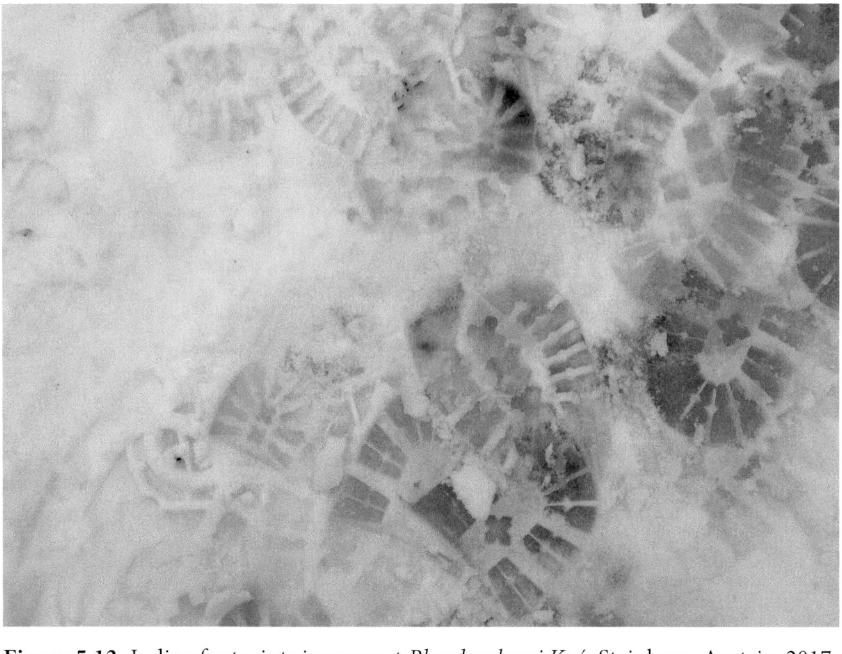

Figure 5.13 Indigo footprints in snow at *Blaudruckerei Koó*, Steinberg, Austria, 2017. Photo: Linda Brassington.

Signposts to the future

In Steinberg, Josef Koó's mangle remains in his family workshop. It extends along the length of the large outhouse as if the walls and roof have been built around it, its chains and pulleys stretching up to the cogs and rollers in the rafters. In Hranovnica, a series of additional outhouses at Montško's workshop bore witness to an earlier practice:

> *A stable, cattle shed, chicken coop, pig sties and rabbit hutches lead to the large majestic wooden barn with its steep pitched roof, ventilation grilles and drying loft above a grey dust floor. This is a space dominated by giant wooden cogs. In former times, pulled by horses, the cogs had driven a massive block of stacked squared timber sections, moving back and forth over heavy rollers bound with tightly wrapped cloth; a 'mangle'.*

Cloth was brought there by women to be pressed ready for block printing, then overdyed in indigo and mangled to a polished finish before taking it back into their daily life as workwear – 'Sunday best' – and household furnishings. The result was smoothly burnished linen, the effect of perpetual motion and friction over layers of compressed bast fibres.

In 1985, in Púchov, Slovakia, a massive mangle resided in the workshop of Stanislav Trnka. Trnka understood the process well, but he had introduced an industrial calendering machine for the polishing of his cloth – the equivalent of a hot iron – and the mangle stood as a monument to the past. In 1985 Trnka had explained the process. His mangle was about 8 metres in length. The complete structure was made of wood. Two rollers sat on a flat bed, weighted down by an enormous series of blocks weighing about 10,000 kilograms (approximately ten tons). Bolted together by crossed metal bars at both ends, the massive structure ran the full length of the bed. Some 30 to 35 metres of cloth were rolled onto one roller. Initially, using only half the pressure, the horse-drawn pulley rotated the cogs, and the mangle pushed the wooden rollers from side to side three times. This action caused the cloth to slacken. At that stage, the cloth was unrolled and rerolled more tightly. The horse circled again and the previous process was repeated, but this time with the full pressure of the wooden structure above. For a third time the cloth was tightened and the wooden structure pushed from side to side three more times. This process relied on the friction of cloth against cloth, fibre against fibre, to produce a unique polished surface, particularly on linen. But by the 1970s the mangle in Púchov was considered surplus to requirements and was sold to the Valašské museum in Rožnov pod Radhoštem, in the

Figures 5.14–5.17 Stages of cloth tensioning and indigo dyeing at *Blaudruckerei Koó*, Steinberg, Austria, 2017. Photo: Linda Brassington.

Figures 5.14–5.17 (*Continued*).

neighbouring Czech Republic, where it is exhibited as a relic of former textile production. Later, the introduction of calendering relied on heated metal rollers, pressing the cloth and compressing the weave; one hard surface against another hard surface, rather than the burnishing of fibres against fibres. Now, in Slovakia, the unique qualities of mangled cloth only appear in historical textiles, distinguishing *modrotlač* ('blueprint') of the early twentieth century from more recent products.

In Steinberg, Koó considers the investment and sacrifice needed to test his mangle – to adjust and maintain its mechanism – is too time-consuming to be cost effective. In addition, Koó's perception that the qualities generated by its process will not be valued has rendered the mangle redundant. Consequently, the mangle in Steinberg stands idle.

In Bhabha's view in *Discussing the Potentials of Cultural Heritage in Austria: Press Conference with Homi K. Bhabha* (2006), the question of heritage presents a 'creative opportunity'. Across Central Europe, the 'mangle' would have been fundamental to past practices of 'blueprint'. As a dye, indigo builds colour through successive immersions and oxidation. Increasing depths of tone result in layers of excess indigo pigment on the surface of the cloth. Its physical properties respond to friction – rubbing, rolling, beating, burnishing, polishing. The compression of linen cloth wound around rollers, and weighed down under giant blocks of solid wood and stones, gave rise to intense lustrous purples, bronzes and blacks. The redevelopment of this monumental device, in parallel with the growing of flax and the recent expansion of linen production, presents a research project for future innovations. In harmony with each other, compression and friction on linen – a fibre with an intrinsic tendency to shine when polished – has the potential to recreate the iridescent surfaces currently unmatched in contemporary 'blueprint'. It is an opportunity to take up the ingenuity of past production methods, to reinvent them, and reapply their fundamental principles in the development of a new material. In so doing, the heritage of 'blueprint' has the capacity to be re-formed into cloth of unique quality – a textile in which the materiality of indigo is central to its design; and substance, rather than imagery, becomes its principal focus.

In the last surviving blueprint workshops of Europe, the successive rhythmical movements between resist paste, printing block and cloth demonstrate an intuitive understanding of craft that goes beyond recipes and techniques. 'Blueprint', as identified in the nations of Austria, the Czech Republic, Germany, Hungary and Slovakia (*Blaudruck, Modrotisk, Kékfestés, Modrotlač: Resist*

Printing and Indigo Dyeing in Europe, 2017 and 2018), expresses a tacit knowledge – a unique judgement, a depth of feeling, an acute visual and tactile sensibility that is both practical and poetic in the process of its making. In 2006, ahead of the international conference *Scope 2, Sites and Subjects: Narrating Heritage*, cultural theorist Homi K. Bhabha considered the definition of heritage. Bhabha suggested it would be more productive to look at heritage in terms of 'provenance', as the 'itinerary of the work' (*Discussing the Potentials of Cultural Heritage in Austria: Press Conference with Homi K. Bhabha*, 2006).

'Blueprint' is a hybrid. Its itinerary reflects its history and development. Since the seventeenth century, itinerant artisans have brought their stories and patterns out of their workshops and into communities. 'Blueprint' has developed through the cross-fertilization of patterns and processes that traversed Central Europe. Wandering craftsmen (journeymen) travelled across regions and territories. They settled in communities, established family businesses and perpetuated the knowledge and understanding of resist printing and indigo dyeing over generations. Despite their rapid decline in the last half of the twentieth century, each 'blueprinter' developed his own distinctive character and figurative design style that have become embedded into regional costume and local customs. As such, dyers and printers have made their own 'interventions', connecting with communities and establishing pattern as part of national identity.

'Blueprint' embodies a strong national identity in its respective European nations, and the resist patterns in its printing blocks hold long traditions and enduring cultural memory. Nevertheless, one might also take the view that their strong traditions have become a barrier to change. How can culture and heritage be acknowledged beyond reproduction and the 'perfect copy'? To perceive the continuation of a process only in textile products is to neglect the significance of other critical contexts – cloth as installation or intervention, in the landscape or the built environment. In Ingold's view:

> Just as to follow a path is to remember the way, so to engage in any practice is . . . to remember how it is done . . . The important thing . . . is that the process should keep on going, not that it should yield precise replicas of past performance. Indeed 'keeping it going' may involve a good measure of creative improvisation. What would really break the continuity, however, would be if people were forcibly constrained to replicate [the past] . . . The effect would be similar to that of a needle becoming stuck in the groove of a record . . . [unable to] keep the music going.
>
> Ingold 2000: 147–8

Figure 5.18 Indigo-dyed cloth oxidizing at *Blaudruckerei Koó*, Steinberg, Austria, 2017. Photo: Linda Brassington.

In May 2018, the exhibition *im.material*, initiated by the Austrian Commission for UNESCO and co-curated by Lisa Niedermayr and Gabriele Detschmann, was the first installation granted permission to exhibit in the dome of the capital's Michaelertor, a heritage site fiercely protected from the potential damage of any physical intervention. The following passage recalls a site visit to Vienna in February 2018:

> *The weight of the city's imperial past seems to bear down beneath the dome of the Hofburg Palace, dominating the cold grey space. The arches of the Michaelertor encapsulate the bitter air into a refrigerated chill. Pedestrians quicken their pace, their heads facing down to protect them from the harsh winds that blow along the passageways.*

Three months later, the heat of summer permeated the stone structure. Bright sunlight cast shadows through the space and the Michaelertor was regenerated into a visitor hub, the gateway to Vienna's cultural history. Lighting technicians and sound engineers appeared at the scene and the *im.material* installation was set in motion.

> *At dusk, intense blue light radiates across the Michaelertor concourse, out through the portals of the building and up through the windows of the dome into the night sky. Sounds emanate from the walls – the clanking of chains moving in and out of large volumes of indigo, the stream of dye slowly diminishing from a surge of falling liquid into a succession of repetitive drips.*

Three international contemporary artists – Linda Brassington, James Muriuki and Lisa Niedermayr – contributed textile works to the installation. In daytime these exhibits, located around the periphery of the dome, served a different purpose. Each piece was the result of a collaboration with the two remaining indigo dyers in Austria – Josef Koó (*Original Blaudruckerei Koó*) and Karl Wagner (*Blaudruck Wagner*) – both recognized for their family heritage and enduring craft production. Through working collaborations in Austria a communication of ideas and a sharing of practices has developed – a fusion of combined knowledge and skills. In *Bringing Things to Life: Creative Entanglements in a World of Materials* (2010), Ingold describes these communications, the exchange of knowledge, as 'networks' or 'meshworks', and he likens a network to netting, 'to woven fabric, the tracery of lace, the plexus of the nervous system or the web of the spider' (Ingold 2010: 11–12). Significantly, Ingold describes this phenomenon, not as lines of interaction, but as lines of *relations*, 'not between but *along*' (2010: 11–12).

The collaboration with *Original Blaudruckerei Koó* was a pivotal point in the developing practice. Access to the Koó facilities enabled the spiralling of 10 metres of cloth around a star frame – a process too heavy for a small studio – and the potential for large-scale work for architectural spaces. Ingold pictures collaboration, the structure of relations, in the shape of a star, 'with a centre surrounded by many radiating lines' (Ingold 2010: 12), a symbol expressed in the revolving star frames of spiralling cloth at the centre of the circular indigo vats in Steinberg. *Textile Installation at Michaelertor: A Collaboration with Original Blaudruckerei Koó* (Brassington 2018) consisted of transparent lengths of indigo-dyed Austrian paper cloth hung in the niches of the passageway, a series of simple colour gradations and tonal variations in contrast to the architectural splendour beneath the dome:

> *The cloths envelop the surrounding light and shadow within their folds, moving gently in the directional flow of air, waiting to capture a momentary glance to draw their audience into their immersive surfaces.*

Figures 5.19–5.20 Blue light installation in *im.material* (Niedermayr, L. and Detschmann, G.) at *Michealertor Kuppel*, Vienna, 2018. Photo: Linda Brassington.

While resident Viennese and visiting tourists bustled through the space of the *Michaelertor*, moving images of dye practices emerged from behind a textile panel – a stitched and pieced work of indigo-dyed cloth by Niedermayr (2018). Blue hand shapes were suspended above – the result of a programme in primary education with Muriuki and Wagner (2018). For Muriuki:

> Sharing work publicly lends itself a life beyond the creator (myself) . . . The work was installed in a culturally significant site . . . accessed for free . . . Many people were able to interact with it . . . work that was not static but also came alive [in] the lighting that accompanied it as well as [in] the gentle breeze that would glide through the building.
>
> Austrian Commission for UNESCO Annual Report 2018

For Niedermayr, 'The blue dome . . . becomes a place of transformation, as a space of transition and . . . direct interaction with passersby. For a short moment of crossing . . . the daily routine can be changed and one can indulge . . . in the blue infinity' (Austrian Commission for UNESCO 2018).

Figures 5.19–5.20 (*Continued*).

im.material was a brief but compelling presentation recorded in the UNESCO report as 'intangible cultural heritage as an artistic intervention' (Austrian Commission for UNESCO 2018). As an exploratory initiative over a period of just five days, it was a step towards the 'transmission of knowledge' (UNESCO 2003c) through contemporary practice, and a celebration of the intangible cultural heritage of indigo and resist dyeing presented in one of Austria's most iconic buildings. The perpetuation of a craft might take many forms unrecognizable from more familiar images of its tradition but, in Bhabha's view, in *Discussing the Potentials of Cultural Heritage in Austria: Press Conference with Homi K. Bhabha* (2006), the possibility of transition is about aspiration, 'not only representing what you think should happen but being aware that your feet are on shifting ground'. What has been critical for this research is the role of contemporary practice in this mission, considering the fine balance between sustainability and reinterpretation, challenging perceptions of indigo and thinking about resist-dye practices in new ways.

Figure 5.21 Linda Brassington (2018). *Textile installation* in *im.material* (Niedermayr, L. and Detschmann, G.) at *Michealertor,* Vienna, Austria, 2018. Photo: Linda Brassington.

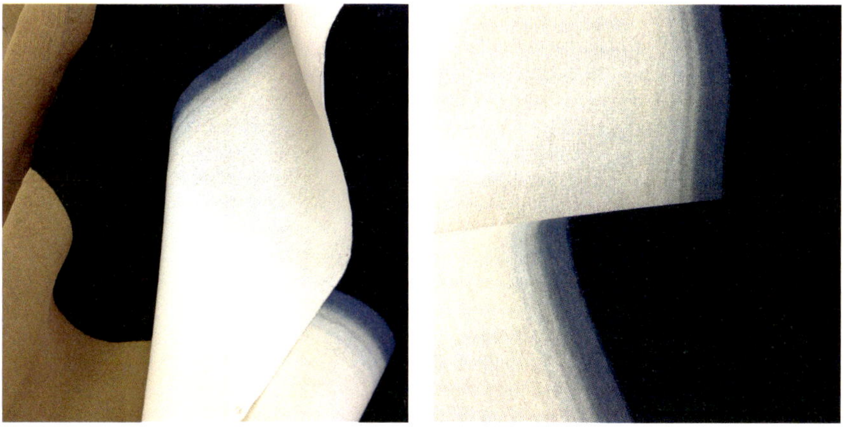

Figures 5.22–5.23 Linda Brassington (2018). Indigo over Austrian paper cloths produced in collaboration with Josef Koó. Photo: Linda Brassington.

Kyoto, Japan

On a hillside in the rural village of Miyama, the late Hiroyuki Shindo worked at his Little Indigo Museum, a 150-year-old building covered in thatch. In 2007 and again in 2019 the exposed eaves revealed the layers of natural material bound together within its roof structure from which a framework of cords, ratchets and winches were suspended above the indigo vats:

> *Buried in compacted earth, grit and straw for more than a century, the indigo vats stand ready for the transformation of colour on cloth. Shindo moves quickly and easily along their edge, lowering and lifting his cloth suspended from the improvised pulley system above. Alongside the workspace, a solitary textile work is displayed at the centre of a raised platform. It is the focus of a single spotlight, a contemplative piece, marking a liminal space between the active and the passive, industry and reflection.*

In his statement accompanying the textile piece *Shindigo 23-01* (2023), exhibited in *Tansa: Japanese Threads of Influence* (2023) at the Crafts Study Centre, Farnham, curated by Lesley Millar, Shindo wrote:

> Half a century has passed since I first became fascinated with indigo. Over that time, through this particular world of indigo, I have learned many things . . . [I]f an artist's job is to bring joy and beauty to people through his or her art, what can art do for us if the process of making art destroys our natural environment?

The piece, created from handwoven cloth of banana and ramie fibre, folded and stacked into a cube with its narrow indigo dyed selvedge, is a testament to

Figure 5.24 Hiroyuki Shindo (2023). *Shindigo 23–01*. Hand woven cloth (banana and ramie fibre) dyed in Japanese indigo. Exhibited at *Tansa: Japanese threads of influence* (Millar, L.), Crafts Study Centre, 2023. Photo: Linda Brassington.

Shindo's values. The compactness of its folds and the focus of its finely coloured edge are an expression of his economy of phrase, his ability to convey the essence of his message for the exhibition as 'an opportunity for both artists [UK and Japanese] to talk about tradition, nature . . .' (*Tansa: Japanese Threads of Influence*, 2023). Four months after Shindo's death in June 2024, while practice continued in his studio, a feeling of emptiness and melancholy emanated through the space. The loom that once accommodated Shindo's most recent work stood idle, as though temporarily interrupted in motion, waiting for its creator to return or for his principles to perpetuate.

In contrast to the workshop of Shindo, weaver Jun Tomita's studio, acquired in 1982, does not reside within a site of heritage. Rather, it has grown out of an abandoned and derelict greenhouse in Koshihata, Kyoto Prefecture, imagined and developed out of a framework of rusted metal from the inside out. A new floor has been constructed for his looms and a new roof withstands the extreme seasonal temperatures of rural Japan. Just as Tomita has created place out of an empty space, only defined by the boundaries of a decaying metal shell, so he has improvised in the building of his equipment – devising a metal scaffold to support the beams, heddles, pedals and reeds of his simple but fully functioning first loom. The following account reflects a momentary experience in 2019 when material substance was allowed to act on its own terms, a passive engagement with dye:

> *Tomita's tensioned warp is stretched down the full length of his glasshouse. Bright daylight pours into the space and onto the painted yarn. As excess dye drips down, loose drops of colour are caught by a cloth placed beneath the threads, and a random scattering of motley speckles are distributed across the floor.*

Tomita has established 'The Village' at his site – a place for the shared practice of his past apprentices, interns and network of artists gathered over the years. He writes, 'I have lived in Koshihata with the desire to convey . . . dyeing and weaving techniques but also the importance of living . . .' (*The Village* 2021), and Shindo's words of tradition and nature seem to echo in Tomita's description of his enterprise.

Mashiko, Japan

Mashiko is the location of Shoji Hamada's development of Soetsu Yanagi's Mingei movement, underpinned by his long association with the potter Bernard Leach and the natural dyer and weaver Ethel Mairet. Higeta is the ninth generation of *katazome* resist dyers in his family. While Mashiko is a ceramics town, the *Higeta Aizome Kobo* is located at its centre, at its crossroads both physically and metaphorically. It is truly at the heart of this place. Its history is longer than the ceramic workshops that have sprung up around it since the last century. In 2016 the dye room presented a space of theatrical scale and proportion:

> *The indigo dye room is situated on one of four sides of a quadrangle, with a courtyard at its centre where the flow of multiple practices crisscross and circulate. Takumi Kojima presides over the operation. The dye workshop is filled with an expanse of seventy-two indigo vats buried in a raised platform of earth – a space set for performance in making. Kojima strides across the multiple rows of wooden lids that cover the vats, knocking and rocking as he moves through the workspace. His rubber boots stomp over the 'stage', treading the boards with confident aplomb. Above the dye vats and the charcoal fires that warm them, the beams and rafters of this 200-year-old building are blackened by woodsmoke, its roof structure impregnated with carbon black. Smoke rises and lingers beneath the vaulted structure, filling the*

Figure 5.25 *Higeta Aizome Kobo* at the crossroads, Mashiko, Japan, 2016. Photo: Linda Brassington.

> *space with a heady atmosphere. Kojima moves over the vats with energy and agility. He crouches down to move his cloth through the dye liquor, stretching it over bamboo frames suspended from the rafters to oxidize its colour, or resting it across trestles perched on the wooden lids of the vats. This place is the site of multiple processing, dynamic movement and lyrical flow determined by its space.*

The story of the *Higeta Aizome Kobo* plays out through its characters – dyers, spinners, weavers – all seen against a backdrop of multiple indigo vats expanding through space. The sense of place at the workshop developed slowly and deliberately over a period of six weeks. In Mashiko, stories of heritage were reflected in two centuries of practice, still held in the memory of Tadashi Higeta himself as the ninth generation of his family: 'When we were kids . . . we'd never go on top of the vats. It had a kind of sacred feel to it . . . so I guess that children somehow felt that sacredness. We didn't have a clear sense of it, but somehow . . . we knew we couldn't go in there' (*Blue Comes from Indigo, Yet it is Bluer than Indigo Itself*, 2024). As such, Tane's 'archaeology of the future' was experienced in the pigmented layers of the workshop's structure and in the compacted earth of the platform.

The art residency in indigo dyeing ran parallel with the exhibition *Inspired by Mairet: Higeta Tadashi and Ethel Mairet* (2016), the first exhibition of textiles at

Figure 5.26 Interior view of *Higeta Aizome Kobo*, Mashiko, Japan, 2016. Photo: Linda Brassington.

Figure 5.27 Indigo vats at *Higeta Aizome Kobo*, Mashiko, Japan, 2016. Photo: Mayumi Tsukimura.

Mashiko Museum of Ceramic Art. The *Higeta Aizome Kobo* carries a long history of *katazome*-resist stencilling demonstrated in the classical Japanese patterns typical of its enterprise. Stencils cut by Tadashi Higeta's father, Hiroshi, are casually pinned on the walls. They are notably more figurative than the repeating patterns evident elsewhere in the workshop. In her essay *Mashiko's Indigo Heritage: Stencil Dyeing by Higeta Hiroshi* (2016), curator Yuko Matsuzaki describes how, following the demise of other indigo workshops during the 1920s and 30s, Hiroshi Higeta was determined to retain the family business. His resolve was triggered by a meeting with Shoji Hamada and his concept of a craft village in Mashiko. Matsuzaki describes how Hiroshi Higeta's stencilled works depict making in Mashiko at that time:

> Hiroshi's *Dyeing Workshop Folding Screen* depicts the indigo dyeing process directly, with the indigo vats sunk into the floor, the stencils placed in a somewhat explanatory manner, the dyed fabric drying, as though looking up at the sky. At first glance, the composition may seem like a casual linking together of sketches from life turned into stencils. When we consider it as a whole, however, we realize that each of the elements has been skilfully combined in a balanced, harmonious composition.
>
> Matsuzaki 2016: 91

According to Matsuzaki (2016), Hiroshi Higeta had said that 'natural colours are more beautiful than the colours people have created' and this perception of dyestuffs is reflected in his work. Matsuzaki writes that 'Higeta Hiroshi unquestionably designed textiles that only he, with his Mashiko roots, could have created' (2016: 91).

Mashiko is located in Tochigi Prefecture of Japan. In 1972, Seiji Ohshima, a noted historian of Western art, was appointed assistant director of the Tochigi Prefectural Museum of Fine Arts and became the museum's director in 1981. Hanging on the wall of the *Higeta Aizome Kobo* is a framed, handwritten document (perhaps a letter) scribed in Japanese by Ohshima:

> That Mashiko is known for 'Mashiko Pottery' goes without saying. But for me, there is one more enjoyment; seeing the indigo-dyeing and hand-weaving work of Mr. Higeta and his son. In Mashiko, which has become increasingly lonely since the death of Dr. Hamada [potter Shoji Hamada], the presence of the Higetas has become irreplaceable to me. When considering the tradition of Mashiko, it may be that Mr. Higeta's indigo dyeing is older than that of the town itself. Mr. Hiroshi Higeta has been silently preserving the old tradition, from the indigo pots to his thatched roof. It must have been quite a hardship to endure the upheaval during and after the war, all while keeping the spark alive in the indigo

> pot and making as well as preserving an enormous number of patterns. He just smiles and says that he is only doing what comes naturally – this is very typical of Hiroshi Higeta. His son, too, looks as if he is simply doing his job, weaving by hand to produce the Moka cotton, which had been on the verge of extinction. The slow rhythm between father and son is natural, and I like it.
>
> Ohshima n.d.

On entering the courtyard on the first morning of the residency, the outdoor space is witnessed as a place of Higeta's work in progress:

> *Tadashi Higeta has spread raw cotton over large bamboo mats slung across trestles. The alternating white, green and brown cotton bolls are grown in a field just a few paces from the workshop. I am encouraged to feel them, the softness of their fibre, their subtle differences. Persimmon fruits, harvested from the tree in the courtyard, are suspended in rows outside, left to ripen in the autumn sunshine, glowing in colour against the dark timbers of the building. Later their juice will yield the tannins for natural colour in Higeta's hand spun cotton, and washing lines are strung across the courtyard with mordanted and dyed skeins of yarn in preparation for weaving.*

Higeta's work is a creative hybrid and his practice has traversed the boundaries of printed and woven textiles. While sustaining his heritage, Higeta has taken a new direction. While colour extracted from his surroundings is central to his work, Higeta has exchanged the repetition of stencils on the surface of the cloth with the rhythm of the loom, weaving colour to create structure and substance. In the essay *Living in the Mashiko Milieu* (2016), curator Satoshi Yokobori describes the significance of local materials in the practices of Mashiko. In the process of making, Hamada had observed that 'If the materials are not indigenous, it does not work' (Yokobori 2016: 31). On his return to Mashiko, after four years of study in woven textiles, Higeta recalled memories of his childhood when local farmers brought brown cotton to the workshop for indigo dyeing. Yokobori writes:

> [Higeta] retained a strong impression of the shades that brown cotton produced. In time, when he inherited the business, he ordered seeds for brown cotton, which had disappeared from Mashiko . . . He himself carried out the work of planting, cultivating, and harvesting as it had been done three decades before. At first, the brown cotton did not seem to work with the Mashiko's clay soil, but over the years the cotton plant accommodated the area's climate and natural features, and their growth improved. He also learned that the colour produced by the brown cotton varied depending on each year's weather.
>
> Yokobori 2016: 32

Significantly, Higeta also learned that 'if he let the cotton rest for three years after harvest, the colour would deepen' – the effect of gradual oxidation in the fibres (Yokobori 2016: 32).

Higeta embodies the knowledge and understanding of tradition while developing independent thinking in his own practice. His work incorporates blends of numerous colours, integrating hues of natural dyed yarns in his woven textiles. All are extracted from his local environment. Intangible cultural heritage is personified in Higeta, not only in his understanding of indigo, but in his sensibility for colour, in the materials and substances of his surroundings, in the colour that is nurtured in his own hands.

Tuan perceives time as motion, as 'flow', and he identifies 'a pause in the temporal current' as the moment when 'space becomes place' (1977: 179). According to Tuan (1975: 165), 'Place is created by human beings for human purposes. A building . . . does not, however, remain a place simply because it is tangible reality and was originally designed as a place. To remain a place it has to be lived in.' But in Tuan's view, this term becomes merely a cliché unless we examine what 'lived in' means: 'To live in a space means to experience it, to be aware of it in the bones, as well as in the head.' As such, 'place, at all scales . . . is a construct of experience; it is sustained not only by timber [and] concrete, but also by the quality of human awareness' (Tuan 1975: 165). During the residency, a series of geometric compositions in *katazome* and indigo was developed in the space:

> *Passing between and across every corner of this heritage site, each group unfolds through a sequence of movement, alternating between rapid activity and moments of stillness. Pausing between each process, the cloths hang passively from their supports – dripping excess water from stainless steel rods, strung up on racks in the drying cabinet, weighted from bamboo canes in the vats, slung from branches above the indigo – all drooping like bats in a state of suspended animation.*

Gradually a series of circles and discs transmuted into seemingly animated ovals, rotating on their axes. Encountering traditions deeply embedded over 200 years, *Pierced* (Brassington 2016) represents a driving force through the residency, a perpetual motion connecting past practices with their future.

UNESCO acknowledges that intangible cultural heritage not only represents inherited traditions from the past but also contemporary practices in which diverse cultural groups take part, sharing expressions of intangible cultural heritage that are similar to those practised by others, 'whether they are from the neighbouring village or from a city on the opposite side of the world' (UNESCO 2003c).

In 2016, during the art residency at the *Higeta Aizome Kobo*, multiple explorations of *katazome* and indigo evolved into a series of transitions from traditional practice into contemporary composition on cloth. Each work related to another, one element of a collective thought, permutations of a single idea. The use of *sasawashi* paper cloth reintroduced Japanese paper fibre in a new context. Its substance, ranging from fine to textured, to heavier weight, determined a variety of dye effects through varying degrees of absorbency. The woven structures gave rise to a range of reversible compositions, revealing the multiple layers of processes in the depth and dimension of their surfaces. In parallel, some imagery was created and printed in England and transported to Japan. This established a foundation upon which to build new layers of colour, tone and composition influenced by the materials and processes particular to Mashiko. The result is a collection of experimental hybrids, responses to space and place through intercultural exchange. The work is not a model of cultural heritage in the form of reproduction. Rather, it represents an exploratory dialogue, a sharing of knowledge and materials, where the results are not on the surface of the work but embedded within it.

Figure 5.28 Gallery talk at the opening of *Linda Brassington: work from the art residency 2016* at Mashiko Museum of Ceramic Art, Mashiko, Japan, April 2017. Photo: Grace Brassington.

Figures 5.29–5.31 Dyed cloth suspended above the indigo vats to oxidize at *Higeta Aizome Kobo*, Mashiko, Japan, 2016. Photo: Linda Brassington.

Figures 5.29–5.31 (*Continued*).

Figures 5.29–5.31 (*Continued*).

Figures 5.32–5.33 The author's *katazome* stencilled cloths suspended from the rafters in preparation for indigo dyeing at *Higeta Aizome Kobo*, Mashiko, Japan, 2016. Photo: Mayumi Tsukimura.

Figures 5.32–5.33 (*Continued*).

Figure 5.34 Linda Brassington (2016). *Diagonal*. Indigo over *katazome* paste resist on discharge printed *sasawashi* paper cloth. Photo: Linda Brassington.

Figure 5.35 Linda Brassington (2016). *Boundary* (reverse). Indigo over *katazome* paste resist on discharge printed *sasawashi* paper cloth. Collection of Mashiko Museum of Ceramic Art. Photo: Linda Brassington.

Figure 5.36 Linda Brassington (2016). *Ellipse 2.* Indigo over *katazome* paste resist on discharge printed *sasawashi* paper cloth. Photo: Linda Brassington.

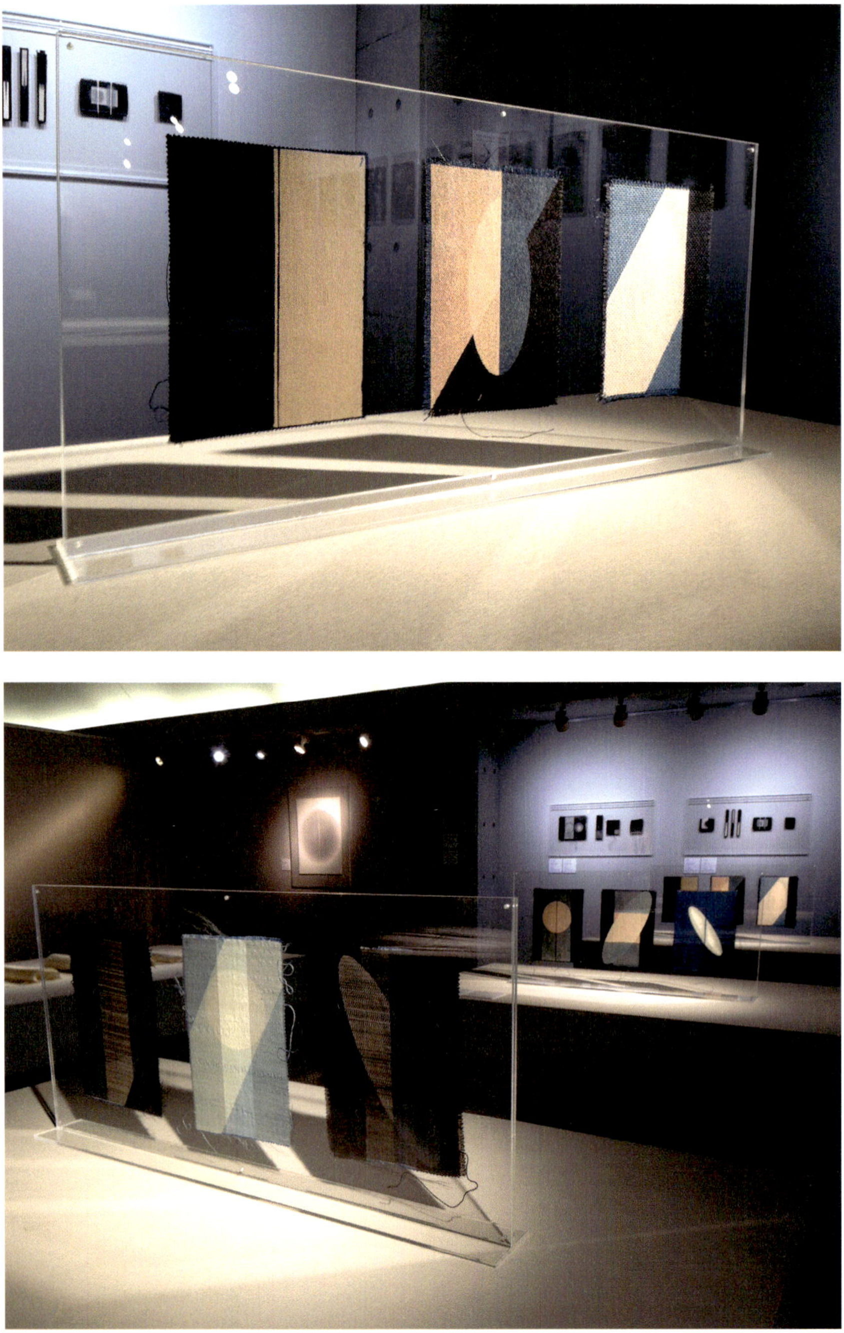

Figures 5.37–5.38 Gallery views of the exhibition *Linda Brassington: work from the art residency 2016* at Mashiko Museum of Ceramic Art, Mashiko, Japan, April–June 2017. Photo: Linda Brassington.

Figure 5.39 Linda Brassington (2016). *Pierced.* Indigo over *katazome* paste resist on discharge printed *sasawashi* paper cloth. Collection of Mashiko Museum of Ceramic Art. Photo: Linda Brassington.

Tokyo, Japan

Fifty miles from Mashiko, in a quiet residential district of Tokyo, the ground breathes the atmosphere of the *Edo Komon* workshop (Komon translates as 'fine patterning' from the Edo period). Here, the Komeya family are masters of *katazome* paste resist, a tangible example of craft heritage. Inside the building, their process depends on an atmosphere of hydration and humidity, qualities essential to retain moisture in the resist paste and to prevent shrinkage of the cloth while stencilling it with a very fine pattern. In 2019 the air is damp. The tangible feeling on cloth, and the intangible sensation in the atmosphere, can only be experienced inside the space. Now, the use of electric humidifiers is employed, but over the 110 years of their practice, the Komeya family have sprayed water onto the earth floor of the workshop at the end of each day.

Third-generation Yasumasa Komeya explains the process. The most important step is done indoors in controlled humidity to prevent the finely-cut stencil from drying out. After first checking the consistency of the resist paste and the condition of the spatula, the stencil is carefully aligned with registration marks (*hoshi*). Then, using a continuous whole-body motion, the resist paste is applied with the exact same pressure from start to finish. The effort demands the utmost concentration.

In the workshop's rafters, rows of numbered planks hewn from finely grained fir are stored above the beams. Each one provides a smooth surface for the spreading of paste onto cloth. The planks are brought down as required and rested on a series of table supports that stand like a sequence of staging posts through the space. After the resist pattern has dried, a dye paste is spread over the whole surface. A sawdust of spruce is scattered over the colour. It has been boiled and dried to remove traces of residual oil and to extract the tannin that might otherwise stain the cloth. Now, the fine wood particles are ready to absorb any excess moisture, while retaining the dampness held in the fibres:

> *This place is a space of light and shadow. The daylight cast through the closed paper screens is subdued, but the extraordinarily fine stencils – the signature of Edo Komon – require light through which to pinpoint the exact position of their repeating image. Rows of electric cables extend along the beams, stretching out above the cloth, casting light from a series of industrial light bulbs at equal intervals along their 7-metre lengths. As the paste is applied, the light bulbs are manoeuvred along the beams. At each point of registration, a spotlight illuminates the stencil, shining light through its cut spaces to ensure an accurate placement of the pattern. The stage is set. As the stencil moves along the narrow cloth, the cable is pulled with it, the light*

> *travelling in tandem with the pattern and moving in rhythm with the hand. The smell of damp lingers in the space, and the paste dries slowly above the earth.*

The compacted mud floor is the embodiment of *Edo Komon*'s heritage in Tokyo. After more than a century, the workshop's dampness continues to emanate through its atmosphere.

In contrast, at the centre of Tokyo's high-fashion district of Aoyama, fabrication is presented as an artform at Homme Plissé Issey Miyake. Ometesando Street is the site of Miyake's flagship store. Interior designer Tokujin Yoshioka has created a retail space with process at its heart. For Yoshioka the appeal of the brand is in the process: '. . . *there* exists its spirit' (Yoshioka cited in Levy 2019). This building of raw concrete and glass has become a theatre of fabric manipulation. Here, pleating and folding is a celebration of process through a ritualistic performance of making. Underneath a smooth white ceiling, a system of industrial extraction ducts is supported against the concrete walls. These mark the space as a place of manufacture. But this is a sealed unit of glass, an animated arthouse, a gallery of construction. In *Dezeen* (Levy 2019), Yoshioka explained his concept: 'I wanted to create a space that shows a documentary-like scene where clothing is created', and in November 2019 its 'spirit' was experienced first-hand:

> *The pleating machine stands on a raised black plinth. It is centred in the space; an icon to process and making. Stage left an artwork fills the wall. It is a print on pleated paper, mounted in a deep black frame behind glass, its image of garment shapes appearing as shadowy traces left behind. The operator enters stage right. He is immaculately dressed in dark blue from head to toe. His uniform consists of a T-shirt under a Miyake vest, pleated cocoon trousers and a peaked cap. Another technician appears donned in a pristine white lab coat with sharp lapels and smooth pockets. He sports sharp creases down the centre of his trousers and his black plimsolls remain spotlessly clean. There is no sign of oil, grease or debris in this sanctified place.*
>
> *As the operator emerges from the sewing room door, he walks to the ironing table. A template is marked out in front of him and he proceeds to press a garment until completely flattened. He works his way along its edges, starting at the centre neck, and working meticulously from the centre out, smoothing and straightening the neckline, shoulder and upper sleeve seams to ensure a sharp geometry. Each sleeve is folded diagonally upwards to create an underarm gusset, the cuffs and wristbands are folded down, and the hems are pressed. Front and back are perfectly aligned, one to the other, and the ritual is repeated. When complete, each garment is carefully laid to one side of the stark black worktop. It is stacked as one of a series of units. A light grey shirt sits alongside them, flat on the surface, its cuffs held down by two silver metal weights.*

Watching and waiting, the garments are fed into the rollers, and the polished stainless steel rods of the pleating machine begin to turn. The perforated bar moves back and forth, pressing the pleats down into the substance of the cloth. The operator presses a series of foil flaps against the cloth, perhaps to compress it or to transmit the heat emanating from the metal rollers. He uses a tool resembling two metal prongs, as if adjusting the pitch of a piano with a tuning fork. Two layers of paper are pushed through the rollers, feeding the garment that is sandwiched between them. At first, scorched papers appear, but the operator moves to the control panel on the right, turning the dials to make a fine adjustment to the temperature.

Slowly the garment emerges from beneath the rollers. The hot papers are cut, swiftly transferred to the worktop, and the layers separated. A pleated cloth is pulled out, its traces of colour left behind on the discarded paper. In a finale to the performance, two T-shaped garments – one in blue and white, the other in black and white – are revealed from between the papers; two printed simulations of Japanese 'shibori'.

Homme Plissé Issey Miyake is identified by its reinterpretations of textile heritage and culture in contemporary fashion. It is a hallowed space, pure and unblemished, an immaculate showcase for Miyake's design vision and technological innovation.

Figures 5.40–5.41 Finished garment at Homme Plissé Issey Miyake, Tokyo, Japan, 2019. Photo: Linda Brassington.

Figures 5.40–5.41 (*Continued*).

Arimatsu, Japan

While Homme Plissé Issey Miyake presents practice as a contemporary drama in a public space, the Harisho enterprise is discreetly located behind the railway line of Arimatsu town. Since 1897, the Harisho family have mastered the traditional resist technique of folding and clamping (*itajime*), and their studio is a heritage site of sculptural fabric forms:

> *The elongated bundles of 30-metre lengths of cloth, compressed into tightly folded squares and triangles, resemble catering packs of sandwiches in their trays. Pins protrude from each package, supporting its manipulations before being clamped in readiness for dyeing. Between each layer is a cut piece of handmade paper, multiple sheets of 'washi' trapped between their folds. Hidden from view, these papers are the secret element, key players in the process of itajime. The sheets of 'washi' serve as buffers between each fold, acting like a wick, and drawing dye into the spaces concealed from view. As the clamps are released, itajime reveals a micro world of resist-dyed patterning on cloth.*

Figure 5.42 Linda Brassington (2021). *Compressed.* Textile, paper, foil. Photo: © Ed Brydon.

At the Harisho studio, much of the action takes place organically in areas obscured from our view. This is exemplified in the traditional technique of *mame*, a pattern identified with Arimatsu's long history of *shibori*. Yoshiko Iwamoto Wada, Mary Kellogg Rice and Jane Barton describe the process as a 'bean pattern', originally created for small towels known as *tenugui* that became a staple product for travellers on their journey along the Tokaido route. According to Wada et al., *mame shibori* was made with 'a special device' (1985: 171). In November 2019, Harisho demonstrated the process.

A 10-metre length of fine cotton is folded into three sections and systematically pushed down into deep grooves cut into the surface of the block. On each block, a strip of wood separates every groove. Each one is protected with a strip of brass hammered onto the surface with fine metal pins. A series of pleated folds ripples over it until the full length of the cloth is taken up. Along each fine strip is a sequence of cut notches. A second block with corresponding grooves is positioned across its opposite form and dovetailed together; horizontal bands over vertical strips. Finally, the two blocks are pushed down until the cloth is fully compressed, tightly clamped together, and locked. At the moment of immersion, the dye floods into the channels where the grooves crisscross over each other, penetrating areas of exposed fibres to create a pattern of multiple repeating *mame* spots on cotton muslin. In *mame*, a pair of interlocking wooden blocks become the point of encounter between cloth and dye in which the action takes place between the clamps and in the grooves of cut wooden boards. Its memory is contained in folded packages – inside the layers of soft cotton cloth, in the intricacies of its substance and the markings of its manipulations. When the unseen spaces are released, *itajime* reveals the micro world trapped between the clamps.

In 2022, the exhibition *Tansa: Japanese Threads of Influence* at the Crafts Study Centre, Farnham, England and at Gallery Gallery, Kyoto, Japan included *Compressed* (Brassington 2021), a miniature work in textile, paper and foil as an imagined place of miniscule scale, an expression of memory and obscured activity hidden within its spaces.

Mashiko revisited

In 2020 the global pandemic resulted in the postponement of a planned return visit to the *Higeta Aizome Kobo* supported by the Great Britain Sasakawa Foundation, the Society of Dyers and Colourists and the Dyers Company, UK. In May 2023, when the work resumed, the return to Mashiko appeared to mark a point of renewed significance. In the post-Covid era, six years after the art residency and subsequent exhibition at the Mashiko Museum of Ceramic Art, the world was questioning the scale of mass production and consumption of textiles. It seemed that the sustainability of Japanese cloth, of which indigo is a part, was of increasing relevance. The Fantasia film production company was visiting the *Higeta Aizome Kobo* to record its processes in the documentary *Blue Comes from Indigo, Yet it is Bluer than Indigo Itself* (2024), and the heritage of Japanese indigo dyeing was positioned to secure a new place in national and global consciousness.

> *This pure, natural indigo gives a clarity of colour, an evenness, a depth, and with the possibility of lighter tones. That's not so easy with synthetic indigo. And the*

> *smell is beautiful, heavy and intoxicating, both in the dye and in the charcoal that warms the vats.*

The focus of the visit was a period of intensive experimentation with indigo and *katazome* over two weeks. A body of new substances was tested – cloth, thread, fibre, organic material – exploring the potential of different fibres, weave structures, reversible effects and combinations of processes. The tonal variations of indigo dye and their proportions of colour were central to every piece. Stretchers of cloth were lowered into vats of different strengths, sometimes at short intervals for pale colour, or for longer periods to achieve intense depth. Tilting the stretchers at different degrees created undulations of dye, and *katazome* paste was applied as a resist to preserve areas of colour through the cloth – a stripe, an edge, a line. In contrast, discharge paste was introduced to strip indigo, to remove areas of colour before overdyeing back into their spaces – building layers on the cloth – on the front and reverse for different effects.

In the documentary film *Blue Comes from Indigo, Yet it is Bluer than Indigo Itself* (2024), Takumi Kojima is shown seated in the courtyard, working swiftly and methodically with his hands. He shreds a pile of short, matted fibres of indigo-dyed hemp, pulling and tearing the fibres apart, collecting the separated substance to be moulded and re-formed on return to the UK. At the *Higeta Aizome Kobo*, ideas and approaches were exchanged:

> *There is a communication through this craft and its material. Even without words, there is an understanding. Just through looking and seeing what's happening on cloth, there is a realization that there is a common language between us.*

Ingold's work *Bringing Things to Life: Creative Entanglements in a World of Materials* (2010) might be seen to express this sharing of practice, this fusion of combined knowledge and skills. A treasured item in a personal collection of material – a compacted coil of vintage thread (ramie or hemp) – became the subject of a new exploratory piece, a metaphor for Ingold's 'creative entanglements' (2010):

> *I rest the twisted spiral of threads, this nest, this network, on suspended mesh and lower it into the indigo . . . slowly . . . As the mass begins to float, bubbles of trapped air gradually rise to the surface . . . popping as they burst. Gently, the threads sink a little lower into the vat. The piece drifts on the surface of the dye . . . for ten to twenty to thirty minutes . . . increasing the depth of tone until the proportions of indigo and undyed space feel right. As the threads begin to disappear beneath the surface, and their buoyancy gives way, the process is abruptly halted. I catch the bundle and draw it up from the vat, the stack is drained, and the fibre is left in air for the indigo to develop its colour.*

Figure 5.43 Linda Brassington (2023). Painted discharge paste on indigo-dyed cloth at *Higeta Aizome Kobo*, Mashiko, Japan, 2023. Photo: Linda Brassington.

The result is a plexus of lines, threads of communication, the point where ideas are bound together at the intersection of material heritage and contemporary practice.

Towards a new definition for intangible cultural heritage

As this exploration has evolved, the overarching experience of practice and field study has become a conduit for the transmission of intangible cultural heritage in indigo and resist dyeing. The direct experience of workspaces past and present, the space and place of resist dyeing in Europe and Japan, has come to redefine

Figure 5.44 Linda Brassington (2023). *Katazome* pasted cloth suspended on bamboo stretchers in indigo at *Higeta Aizome Kobe*, Mashiko, Japan. 2023. Photo: Linda Brassington.

intangible cultural heritage. In these theatres of making, the intangible has been made tangible in another form, the reconfiguration of traditional craft in contemporary practice.

The remains of Hranovnica in Slovakia are essential in holding the memory of Slovak 'blueprint' within its spaces. To separate the processes and practices of resist dyeing from their physical environment would be to diminish their meaning and authenticity. In parallel, the significance of international collaboration and exchange in Central Europe and Japan, the lines of communication they have created and the network of ideas they have generated, express Bhabha's notion of

Figure 5.45 Linda Brassington (2023). Indigo dyed hemp fibre at *Higeta Aizome Kobo*, Mashiko, Japan, 2023. Photo: Linda Brassington.

heritage as a 'creative opportunity' (*Discussing the Potentials of Cultural Heritage in Austria: Press Conference with Homi K. Bhabha*, 2006).

This book has set out a new space for intangible cultural heritage. Through the exploration of metaphorical and performance meanings in resist dyeing, this exploration has moved towards a new critique for the classification of Intangible Cultural Heritage. It has stretched the boundaries of UNESCO's identification, redefining it beyond skills and knowledge, and into a sensory space of lived experience – a place for practice and continuing dialogue.

Figure 5.46 Linda Brassington (2023). *Entangled*. Ramie in indigo at *Higeta Aizome Kobo*, Mashiko, Japan, 2023. Photo: Linda Brassington.

Artworks

Barney, M. (2015), *Water Castings* (sculpture, bronze), Los Angeles: Museum of Contemporary Art, 13 September 2015 to 18 January 2016.

Brassington, L. (*c.*1985), *Resist dyed length* (textile, indigo, myrabolan and iron rust over hand-drawn waxed lines), Farnham: Crafts Study Centre Collection, ID Number Current Accession 2002: 3.

Brassington, L. (2016), *Arc 1* (textile, *katazome* paste resist and discharge print on *sasawashi* paper cloth with foil), Mashiko: Mashiko Museum of Ceramic Art, 23 April to 2 July 2017.

Brassington, L. (2016), *Arc 2* (textile, indigo over *katazome* paste resist on discharge printed *sasawashi* paper cloth with foil), Mashiko: Mashiko Museum of Ceramic Art, 23 April to 2 July 2017.

Brassington, L. (2016), *Boundary* (textile, indigo over *katazome* paste resist on discharge printed 'sasawashi' paper cloth with foil), Mashiko: Mashiko Museum of Ceramic Art, 23 April to 2 July 2017.

Brassington, L. (2016), *Diagonal* (textile, indigo over *katazome* paste resist on discharge printed *sasawashi* paper cloth), Mashiko: Mashiko Museum of Ceramic Art, 23 April to 2 July 2017.

Brassington, L. (2016), *Ellipse 1* (textile, indigo over *katazome* paste resist on discharge printed *sasawashi* paper cloth), Mashiko: Mashiko Museum of Ceramic Art, 23 April to 2 July 2017.

Brassington, L. (2016), *Ellipse 2* (textile, indigo over *katazome* paste resist on discharge printed *sasawashi* paper cloth), Mashiko: Mashiko Museum of Ceramic Art, 23 April to 2 July 2017.

Brassington, L. (2016), *Encrusted* (textile, indigo, pumice and acrylic on Japanese linen cloth), Mashiko: Mashiko Museum of Ceramic Art, 23 April to 2 July 2017.

Brassington, L. (2016), *Pierced* (textile, indigo over *katazome* paste resist on discharge printed *sasawashi* paper cloth), Mashiko: Mashiko Museum of Ceramic Art Collection (2016).

Brassington, L. (2016), *Torn 2* (textile, indigo over *katazome* paste resist on discharge printed *sasawashi* paper cloth with foil), Mashiko: Mashiko Museum of Ceramic Art, 23 April to 2 July 2017.

Brassington, L. (2017), *Rhythm* (textile, indigo over *katazome* paste resist on linen), Mashiko: Mashiko Museum of Ceramic Art, 23 April to 2 July 2017.

Brassington, L. (2018), *Textile installation at Michaelertor: a collaboration with Original Blaudruckerei Koó* (textile, indigo and paper cloth), Vienna: Hofburg Palace, 9 to 13 May 2018.

Brassington, L. (2021), *Compressed* (textile, paper, foil), Farnham: Crafts Study Centre, 4 January to 26 March 2022; Kyoto: Gallery Gallery, 23 April to 8 May 2022.

Brassington, L. (2023), *Entangled* (indigo on ramie).

Bosence, S. (1983), *Resist patterned length* (textile, vat dye over resist paste on indigo-dyed cotton), Farnham: University for the Creative Arts, Textiles Collection. ID Number Current Accession 7943.

Fukumoto, S (1996), *Constellation* (textile), Osaka: Artcourt Gallery.

Fukumoto, S. (1996), *Tea Room – Mist* (installation, textile), Nogata: Nogata Small and Medium Enterprises University.

Fukumoto, S. (2001), *Untitled* (textile, indigo dyed cotton sateen over stitch resist), Farnham: Collection of Lesley Millar.

Gallaccio, A. (1999/2018), *All the Rest is Silence* (installation, textile), Southampton: John Hansard Gallery, 24 November 2018 to 2 February 2019.

Hiorns, R. (2008/2013), *Seizure* (intervention, copper sulphate), London: Artangel/ Jerwood, 151–189 Harper Road, 3 September to 30 November 2008; Wakefield: Yorkshire Sculpture Park, Arts Council Collection, 2013.

Kapoor, A. (1990), *Void* (fibreglass and pigment), London: Royal Academy of Arts. 26 September to 11 December 2009.

Kapoor, A. (1992), *Artist's sketchbook.*

Long, R. (1993), *Dusty to Muddy to Windy* (textwork), Bristol: Arnolfini, 31 July to 15 November 2015.

Long, R. (2011), *Avon River Mud Circle* (installation, mud), Bristol: M-Shed.

Long, R. (2014), *Untitled* (diptych, china clay on plywood).

Nai, M. (2018). *Untitled* (installation series, indigo dyed jute, wood), Miami Beach: Art Basel, 2019.

Shindo, H. (1995), *Thread Ball* (textile, linen, cotton, indigo dye), Wilton, CT: Browngrotta Arts.

Shindo, H (2023), *Shindigo 23-01* (textile, banana fibre, ramie, indigo dye), Farnham: Crafts Study Centre, 4 January to 26 March 2022; Kyoto: Gallery Gallery, 23 April to 8 May 2022.

Shiota, C. (2009), *Inside – Outside* (installation, recycled wood), Berlin: Hoffmann Collection.

Shiota, C. (2010), *Farther Memory* (installation, recycled wood), Teshima: Art Setouchi.

Shiota, C. and T. Tane (2019–22), *Teshima Project: In Progress* (concept models, images and objects), Teshima: Art Setouchi.

Sudo, R. (2019), *Jellyfish* (textile), Hong Kong: CHAT, the Centre of Heritage, Arts and Textile, 23 November 2019 to 1 March 2020.

Sudo, R. (2019), *Koi Currents* (installation, textile), Hong Kong: CHAT, the Centre of Heritage, Arts and Textile, 23 November 2019 to 1 March 2020.

Sudo, R. (2020), *Ogi No Mai / Japanese Fanfare* (installation, textile), Compton Verney: Compton Verney Art Gallery, 14 March to 5 November 2020.

Tanaka, C. (1991), *Grinded Fabric – Ocher* (textile, ochre rubbed over handwoven ramie and indigo-dyed warp), Collection of Lesley Millar.

Tanaka, C. (1994), *Six Squares, Indigo Blue* (textile), Farnham: James Hockey Gallery, 27 April to 23 June 2001.

Tane, T. (2012), *Kofun Stadium: New National Stadium of Japan* (architectural design), Paris: Atelier Tsuyoshi Tane Architects.

Wilson, R. (1987/2018), *20:50*, London: Hayward Gallery, 26 September 2018 to 6 September 2019.

References

Anish Kapoor (2009), exhibition, London: Royal Academy, 26 September to 11 December 2009.

Anish Kapoor: Ann Lewis AO Contemporary Visual Arts International Address (2012), online video, Melbourne: Australian Museum of Contemporary Art, 20 January 2012, https://www.youtube.com/watch?v=4Mt3fGklf4A.

Anya Gallacio: All the Rest is Silence (1999–2018) (2018), press release, Southampton: John Hansard Gallery, http://www.jhg.art/event-detail/350-anya-gallaccio-all-the-rest-is-silence-/.

Artangel (2020), webpage, *About Us*, https://www.artangel.org.uk/about_us/.

Atelier Tsuyoshi Tane Architects (2019), website, France: http://at-ta.fr.

Auden, W. H. (1944), *The Sea and The Mirror: A Commentary on Shakespeare's The Tempest*, https://books.google.co.uk/books?hl=en&lr=&id=eNfmdU4m4TYC&oi=fnd&pg=250PR7&dq=all+the+rest+is+silence+wh+auden&ots=MBLtEKSeXn&sig=Obr6mc9mOFjo3bArWucNFpKx5jg#v=onepage&q&f=false.

Austrian Commission for UNESCO (2017), *Blaudruck/Modrotisk/Kékfestés/Modrotlač: Resist Block-printing and Indigo-dyeing in Europe*, Nomination to the Representative List of Intangible Cultural Heritage, Vienna: Austrian Commission for UNESCO.

Austrian Commission for UNESCO (2018), 'Intangible cultural heritage as an artistic intervention: Blueprint project *im.material* – Light installations and exhibition in public space', *Annual Report 2018*, 29–31, Vienna: Austrian Commission for UNESCO.

Balfour-Paul, J. (1997), *Indigo in the Arab World*, Richmond: Curzon Press.

Balfour-Paul, J. (1998), *Indigo: Egyptian Mummies to Blue Jeans*, London: British Museum Press.

Balfour-Paul, J. (2015), *Deeper than Indigo: Tracing Thomas Machell, Forgotten Explorer*, London: Medina.

Bancroft, E. (1813), *The Philosophy of Permanent Colours*, London: Cadell and Davies.

Barnes, R. (1997a), 'Introduction: the collection', in *Indian Block-Printed Textiles in Egypt: The Newberry Collection in the Ashmolean Museum, Oxford*, http://www.jameelcentre.ashmolean.org/collection/7/10236/10316.

Barnes, R. (1997b), 'Resist- and mordant-dyeing', in *Indian Block-Printed Textiles in Egypt: The Newberry Collection in the Ashmolean Museum, Oxford*, http://www.jameelcentre.ashmolean.org/collection/7/10236/10321.

Barron, P. (1964/2018), 'My Life as a Block Printer', Crafts Study Centre, in M. Silver and S. Burns (eds), *Barron and Larcher: Textile Designers*, 42–80, Woodbridge: ACC Art Books.

Baume, N., ed. (2008), 'Mythologies in the making: Anish Kapoor in conversation with Nicholas Baume', in *Anish Kapoor: Past, Present, Future*, Boston: Institute of Contemporary Art.

Belsey, F. (2014), *Miao Textile Collections at the Pitt Rivers Museum*, https://www.academia.edu/32375626/Maio_Textile_Collections_at_the_Pitt_Rivers_Museum_docx?auto=download.

Bhabha, H. (2009a), 'Anish Kapoor: Making Emptiness', in A. Kapoor, H. Bhabha, J. de Loisy et al., *Anish Kapoor*, 171–7, London: Royal Academy of Arts.

Bhabha, H. (2009b), 'Elusive Objects: Anish Kapoor's Fissionary Art', in A. Kapoor, H. Bhabha, J. de Loisy et al., *Anish Kapoor*, 24–35, London: Royal Academy of Arts.

Blake, W. (1863), *Auguries of Innocence*, https://www.poetryfoundation.org/poems/43650/auguries-of-innocence.

Blaudruck Wagner: Traditionelle Handwerkskunst seit 1878, website, http://www.blaudruck.at.

Blaudruck/Modrotisk/Kékfestés/Modrotlač, Resist Block Printing and Indigo Dyeing in Europe (2017), online video, Austria: Austrian Commission for UNESCO, 28 November 2018, https://www.youtube.com/watch?v=Q-L0phapf80&t=24s.

Blaudruck/Modrotisk/Kékfestés/Modrotlač, Resist Block Printing and Indigo Dyeing in Europe (2018), webpage, UNESCO: Representative List of the Intangible Cultural Heritage of Humanity, https://ich.unesco.org/en/RL/blaudruck-modrotisk-kekfestes-modrotlac-resist-block-printing-and-indigo-dyeing-in-europe-01365.

Blue Comes from Indigo, Yet it is Bluer than Indigo Itself (2024), film, dir. S. Yoshida, trans. R. Ohkanda, Japan: Fantasia.

Bosence, S. (1978), 'Technical notes and glossary', in B. Roscoe and S. Bosence (eds), *Hand Block Printing: Barron and Larcher 18 slides*, 7–11, London: Crafts Council.

Bosence, S. (1985), *Hand Block Printing and Resist Dyeing*, Newton Abbot: David and Charles.

Bosence, D. (1996), 'Conclusions and thanks', in *Words and Music: A Celebration of the Life and Work of Susan Bosence, The Great Hall, Dartington, 16 June 1996*, 23–4, Dartington: Dartington Trust.

Brassington, L. (1987), *Modrotlač: Indigo Country Cloths and Artefacts from Czechoslovakia*, Farnham: West Surrey College of Art and Design.

Brassington, L. (2004), *The Collection as a Narrative in the Crafts*, Farnham: Crafts Study Centre, https://www.csc.uca.ac.uk.

Brassington, L. (2006), 'The Muriel Rose Bequest', in J. Vacher (ed.), *Muriel Rose: A Modern Crafts Legacy*, 42–55, Farnham: Crafts Study Centre.

Brassington, L. (2015), 'Immersed in Indigo. A test of faith: shibori and the work of Jane Callender', *Selvedge* 63: 62–5.

Brezinski, D., ed. (2016), *Super-black Coatings*, blogpost, USA: *Coatings World*, 4 November 2016, https://www.coatingsworld.com/contents/view_blog/2016-04-11/super-black-coatings/.

Buhler, A., ed. (1983), *Block Printing and Dyeing of Bagru, Rajasthan*, Ahmedabad: Calico Museum of Textiles.
Burns, S., N. Crosland and L. Loakes (2018), 'The Printmakers' Perspective', in M. Silver and S. Burns (eds), *Barron and Larcher: Textile Designers*, 95–115, London: Christopher Farr Cloth.
Butler, R. (2008), 'In the Mud with Richard Long: the land artist who uses mud from the River Avon', *1843, The Economist, Autumn 2008,* https://www.1843magazine.com/story/mud-with-richard-long.
Callender, J. (2017) *Stitched Shibori: Technique, Innovation, Pattern, Design*, Tunbridge Wells: Search Press.
Cardon, D. (2007), *Natural Dyes: Sources, Traditions, Technology and Science*, London: Archetype.
Cardon, D., ed. (2016), *The Dyer's Handbook: Memoirs of an 18th Century Master Colourist*, Oxford and Havertown: Oxbow Books.
Čechmánková, N. (2012), *Josef Vydra as we know and do not know him*, http://www.uluv.sk/en/web/magazine/archive/year-2012/rud-042012/josef-vydra-as-we-know-and-do-not-know-him/.
CHAT Winter Programme (2019), online video, Hong Kong: CHAT Centre for Heritage, Arts and Textile, https://www.youtube.com/watch?v=e0-Cs0P3 yw&feature=emb_logo.
Chiyoko Tanaka (2024), webpage, USA: Browngrotta Arts, https://browngrotta.com/artists/chiyoko-tanaka.
Coatts, M. (1996), 'Obituaries: Susan Bosence', *The Independent*, 21 February 1996, https://www.independent.co.uk/news/people/obituaries-susan-bosence-1320127.html.
Crafts Council and authors (1994), *Colour into Cloth*, London: Crafts Council.
Crafts Study Centre Collections, https://vads.ac.uk/digital/collection/CSC.
Cresswell, T. (2004), *Place: A Short Introduction*, Oxford: Blackwell.
Daiwa Anglo-Japanese Foundation (2004), *Expressions in Indigo: The Textile Art of Fukumoto Shihoko*, exhibition webpage, London: Daiwa Anglo-Japanese Foundation, 8 September to 22 October 2004, https://dajf.org.uk/exhibitions/expressions-in-indigo-the-textile-art-of-fukumoto-shihoko.
Danglová, O. (2014), *Blueprint in Slovakia*, Bratislava: ÚĽUV Centre for Folk Art Production.
De Waal, E. (2010), *The Hare with Amber Eyes: A Hidden Inheritance*, London: Chatto and Windus.
Delaney, B. (2016), '"You could disappear in it": Anish Kapoor on his exclusive rights to the "blackest black"', *Guardian*, 26 September 2016, https://www.theguardian.com/artanddesign/2016/sep/26/anish-kapoor-vantablack-art-architecture-exclusive-rights-to-the-blackest-black.
Discussing the Potentials of Cultural Heritage in Austria: Press Conference with Homi K. Bhabha (2006), online video, UNESCO Scope 2 Sites and Subjects: Narrating Heritage conference, Vienna: 12 July 2006, https://www.youtube.com/watch?v=8AhDyvl08_Q.

Dyeing Now: Contemporary Makers Celebrate Ethel Mairet's Legacy (2016), webpage, Ditchling Museum of Art and Craft, https://www.ditchlingmuseumartcraft.org.uk/event/dyeing-now/.

The Exhibition of the Works by Linda Brassington (2017), exhibition, Japan: Mashiko Museum of Ceramic Art, 23 April to 2 July 2017.

Fabric: Touch and Identity (2020), exhibition, UK: Compton Verney, 14 March to 5 November 2020, https://www.comptonverney.org.uk/thing-to-do/fabric-touch-and-identity/.

Farther Memory (2013), webpage, Japan: Setouchi Explorer, https://www.setouchiexplorer.com/farther-memory/.

Faulkner R. (2010), 'Ken Mihara and Shihoko Fukumoto: Ceramics and Indigo-dyed Textiles', *Galerie Besson*, http://www.galeriebesson.co.uk/miharafukumoto.html.

Frieze Masters 2013: Richard Long (2013), online video, London: Lisson Gallery, 15 October 2013, http://www.lissongallery.com/news/frieze-masters-2013-richard-long.

Fukumoto Shihoko: Japan Blue (2015), exhibition webpage, USA: Artcourt Gallery, https://www.artcourtgallery.com/eng/exhibitions/5497/.

Fukumoto, S. (2016), *Urges/Urgings: Notes on a Theory of the Art of Japanese Dyeing*, Osaka: Osaka University of Arts.

Garfield, S. (2000), *Mauve: How One Man Invented a Colour that Changed the World*, London: Faber and Faber.

Gayford, M. (2018), 'There's almost nothing in this Hayward show – and that's the point', *Spectator*, https://www.spectator.co.uk/2018/09/theres-almost-nothing-in-this-hayward-show-and-thats-the-point/.

Gittinger, M. (1982), *Master Dyers of the World*, Washington, DC: The Textile Museum.

Granger-Taylor, H. and D. O'Connor (1982), *Colour and the Calico Printer: An Exhibition of Printed and Dyed Textiles 1750–1850*, Farnham: West Surrey College of Art and Design.

Handmade in Japan: Mud Dyeing the Silk (2017), TV programme, BBC4, https://www.bbc.co.uk/programmes/p055ypkh.

Hara, K. (2007), *White*, Zurich: Lars Müller Publications.

Hiroyuki Shindo (2017), webpage, USA: Browngrotta Arts, http://www.browngrotta.com/Pages/shindo.php.

im.material (2018), exhibition, UNESCO Austria, Vienna: Michaelertor, 9 May to 13 May 2013.

In Conversation: Lesley Millar and Reiko Sudo (2020), online talk, UK: Compton Verney: Compton Verney Art Gallery, 10 December 2020.

In Conversation: Matthew Barney and Homi K. Bhabha (2015), online video, Los Angeles: Museum of Contemporary Art, 11 October 2015, https://www.youtube.com/watch?v=7M0IFSJjFiM.

Ingold, T. (2000), *The Perception of the Environment: Essays on Livelihood, Dwelling and Skill*, London: Routledge.

Ingold, T. (2006), 'Up, across and along', *Place and Location: Studies in Environmental Aesthetics and Semiotics* 5: 21–36.

Ingold, T. (2007), 'Materials against materiality', *Archaeological Dialogues* 14, no. 1: 1–16.

Ingold, T. (2009a), 'The Display of Creative Practice', lecture, Department of Media and Culture Studies, Utrecht University and the Meertens Institute, Royal Netherlands Academy of Arts and Sciences (eds), *Showing Making: International Conference on the Representation of Image Making and Creative Practices in Ritual, Art, Media and Science Conference Proceedings, Film Museum, Amsterdam, 18 June to 19 June 2009*, Amsterdam: Dutch Utrecht University, Royal Netherlands Academy of Arts and Sciences, Dutch Filmmuseum.

Ingold, T. (2009b), 'The textility of making', *Cambridge Journal of Economics* 34, no. 1: 91–102.

Ingold, T. (2010), 'Bringing Things to Life: Creative Entanglements in a World of Materials', *ESRC National Centre for Research Methods NCRM Working Paper*, Series 5, no. 10: 1–16.

Ingold, T. (2017a), *Correspondences: Knowing from the Inside*, Aberdeen: University of Aberdeen.

Ingold, T. (2017b), 'Surface Visions', *Theory, Culture and Society, University of Aberdeen* 34, no. 7–8: 99–108.

Ingold, T. (2017c), 'What if the city were an ocean and its buildings ships?', in S. V. Bradley (ed.), *The Evergreen: A New Season in the North* 2: 1–5.

Ingold, T. (2019a), *In the Gathering Shadows of Material Things*, lecture, London: Department of History of Art, University College London, 7 February 2019.

Ingold, T. (2019b), *What on Earth is the Ground?*, lecture, London: University of the Arts London, Central St Martins, 9 April 2019.

Inspired by Mairet: Higeta Tadashi and Ethel Mairet (2016), exhibition, Japan: Mashiko Museum of Ceramic Art, 30 October 2016 to 29 January 2017.

International Biennial of Tapestry (1990), exhibition, Lausanne: Musée Cantonal des Beaux-Arts.

Interview: Roger Hiorns on Seizure (2008), online video, UK: Artangel, https://vimeo.com/7184407.

Jacob, M. J. (1993), *Shihoko Fukumoto: Shades of Indigo Blue*, Takashimaya: Gallery at Takashimaya.

Jane Weir, webpage, UK: Poetry Archive, https://poetryarchive.org/poet/jane-weir/.

TheJapanPage.com, website, https://www.thejapanpage.com.

Jeffries, S. (2015), 'Thomas Machell: the explorer, the cannibal and me', *Guardian*, 3 June 2015, https://www.theguardian.com/artanddesign/2015/jun/03/thomas-machell-indigo-jenny-balfour-paul-interview.

John Hansard Gallery (2018), 'Anya Gallacio: All the Rest is Silence', Twitter post, https://twitter.com/hashtag/JHGGallaccio?src=hash.

Jones, J. (2018), 'Richard Long review – Modern Primitive sees the Cosmos Reflected in Mud', *Guardian*, https://www.theguardian.com/artanddesign/2018/may/11/richard-long-review-circle-to-circle-mud-sun-moon-stars.

Kennedy, R. (2015), *Matthew Barney's Most Punishing Tour: 'River of Fundament'*, https://www.nytimes.com/2015/09/20/arts/design/matthew-barneys-most-punishing-tour-river-of-fundament.html.

King, B. (2019), *The Wardle Family and its Circle: Textile Production in the Arts and Crafts Era*, Woodbridge: Boydell Press.

Kofun Stadium (2012), design project, Atelier Tsuyoshi Tane Architects, http://at-ta.fr.

Levy, N. (2019), 'Tokujin Yoshioka puts Garment Making on Show Inside Tokyo's Homme Plisse Flagship', *Dezeen*, 6 August 2019, https://www.dezeen.com/2019/08/06/homme-plisse-flagship-store-interiors-tokyo-issey-miyake/.

The Lost Textile of Ryukyu (2020), TV programme, Japan: NHK World Prime, 7 November 2020, https://www3.nhk.or.jp/nhkworld/en/tv/worldprime/20201107/3004681/.

Mahtani, N. (2018), 'Bombay Blues: Manish Nai Takes us on an Inky Trail of the Maximum City', *Architectural Digest India*, 19 November 2018, https://www.architecturaldigest.in/content/manish-nai-mumbai-blue/.

Mailer, N. (1983), *Ancient Evenings*, Boston: Little, Brown and Company.

Mairet, E. M. (1916) *A Book of Vegetable Dyes: A Book of Recipes and Information Useful to the Dyer*, Ditching: Ditchling Press.

Martin, C. (2019), *January 23, 2019*, Twitter post, https://twitter.com/Chris_M_Photo/status/1088085998297366528.

Marx, E. (1985), 'Working with Barron and Larcher', in S. Bosence, *Hand Block Printing and Resist Dyeing*, 131, Newton Abbott: David and Charles.

Matsuzaki, Y. (2016), 'Mashiko's Indigo Heritage: Stencil Dyeing by Higeta Hiroshi', in *Inspired by Mairet: Higeta Tadashi and Ethel Mairet*, 90, Mashiko: Committee for Promotion of Mashiko Culture Town.

Matthew Barney: River of Fundament (2015), exhibition, USA: Museum of Contemporary Art in Los Angeles, 13 September 2015 to 18 January 2016.

McCarty, C. and M. McQuaid (1998), *Structure and Surface: Contemporary Japanese Textiles*, New York: Museum of Modern Art.

McQuaid, M. (1998), 'Transformations: Process and Technique in Contemporary Japanese Textiles', in C. McCarty and M. McQuaid, *Structure and Surface: Contemporary Japanese Textiles*, 17–31, New York: Museum of Modern Art.

Millar, L. (2001), *Textural Space: Contemporary Japanese Textile Art*, Farnham: Surrey Institute of Art and Design, University College.

Millar, L., T. Uchiyama and C. Tanaka (2002), *Chiyoko Tanaka: Portfolio Collection 12*, Winchester: Telos Art.

Modrotlač (2014), exhibition, Bratislava: ÚĽUV The Centre for Folk Art Production, 2014 to 2015.

Modrotlač Peter Trnka, website, http://www.modrotlac.sk.

Moreton, G. (2018), 'November 13, 2018', Twitter post, https://twitter.com/guy_moreton/status/1068414230746079232.

Morris, W. (1876a/1988), 'Indigo dyeing with Wardle', letter to Aglaia Coronio, 28 March 1876, in G. Naylor (ed.), *William Morris by Himself: Designs and Writings*, London: Macdonald and Co.

Morris, W. (1876b/1988), 'Transferring indigo production from Leek to Hammersmith', letter to Thomas Wardle, 9 May 1876, in G. Naylor (ed.), *William Morris by Himself: Designs and Writings*, London: Macdonald and Co.

Morris, W. (1877a/1988), 'Indigo dyeing with Wardle', letter to Georgiana Burne-Jones, 4 February 1877, in G. Naylor (ed.), *William Morris by Himself: Designs and Writings*, London: Macdonald and Co.

Morris, W. (1877b/1988), 'Indigo dyeing with Wardle', letter to Janey Morris, 7 February 1877, in G. Naylor (ed.), *William Morris by Himself: Designs and Writings*, London: Macdonald and Co.

Morris, W. (1877c/1988), 'Transferring indigo production from Leek to Hammersmith', letter to Thomas Wardle, 30 November 1877, in G. Naylor (ed.), *William Morris by Himself: Designs and Writings*, London: Macdonald and Co.

Nakano, E. (1982), *Japanese Stencil Dyeing: Paste-Resist Techniques*, New York: Weatherhill.

Naylor, G., ed. (1988), *William Morris by Himself: Designs and Writings*, London: Macdonald and Co.

O'Connor, D. (1982), 'The Work is Part of Life', *Crafts* 56, no. 3: 42–8.

O'Connor, D. (1985a), 'Foreword', in S. Bosence, *Hand Block Printing and Resist Dyeing*, Newton Abbot: David and Charles.

O'Connor, D. (1985b), 'Technical Notes for Wax', teaching materials, Farnham: West Surrey College of Art and Design.

O'Connor, D. (1993), 'Re: Record of observations in China', personal notebook, Farnham: The Textiles Collection, University for the Creative Arts.

O'Connor, D. (1994a), *Miao Costumes*, Farnham: James Hockey Gallery.

O'Connor, D. (1994b), 'Re: Record of observations in China', personal notebook, Farnham: The Textiles Collection, University for the Creative Arts.

O'Connor, D. (2009/2014), 'Little Flowered Miao', in F. Belsey, *Miao Textile Collections at the Pitt Rivers Museum*, https://www.academia.edu/32375626/Maio_Textile_Collections_at_the_Pitt_Rivers_Museum_docx?auto=download.

Ohshima, S. (n.d.), handwritten document, trans. R. Ohkanda.

Olding, S. (2012), 'Craft Studio', in S. Murison-Bowie (ed.), *Artists and Studios: Private Views*, 10–15, Oxford: Oxford Folio.

Original Indigo Blaudruckerei Koó, website, http://www.originalblaudruck.at.

Picton, J. and J. Mack (1989), *African Textiles: Looms, Weaving and Design*, London: British Museum Press.

Podoba, J. (1988), 'Buildings and utilization of space in the blueprint workshop in Hranovnica', *Bulletin of the State Institute of Monument Preservation, Bratislava*: 101–6.

Powers, A. (1992), *Modern Block Printed Textiles*, London: Walker Books.
Richard Long: Circle to Circle (2018), exhibition, London: Lisson Gallery, 11 May 2018 to 23 June 2018.
Richard Wilson at Space Shifters: In the Gallery (2018), online video, London: Southbank Centre, https://www.youtube.com/watch?v=Zg4E4o96Qjs.
Robinson, S. (1969a), *A History of Printed Textiles*, London: Studio Vista.
Robinson, S. (1969b), *A History of Dyed Textiles*, London: Studio Vista.
Robinson, S. (1971), 'The Development of the Cotton Industries: Chintz 1650–1800', in *A History of Dyed and Printed Textiles*, London: Visual Publications.
Roger Hiorns, Seizure 2008/2013 at Yorkshire Sculpture Park (2013), online video, England: Arts Council Collection, 18 June 2013, https://www.youtube.com/watch?v=PoRyvpfkc1U&t=51s.
Ronald, E. (2007) *Balotra: The Complex Language of Print*, Jaipur: Anokhi Museum of Hand Printing, 60–1.
Roscoe, B. (1978), 'Hand Block Printing: Barron and Larcher', in B. Roscoe and S. Bosence (eds), *Hand Block Printing: Barron and Larcher 18 slides*, 1–6, London: Crafts Council.
Roscoe, B. (1994), 'Susan Bosence: an interview with Barley Roscoe', in Crafts Council and authors (eds), *Colour into Cloth*, London: Crafts Council.
Royal Economic Society (1912), 'The Rise and Fall of the Indigo Industry in India', *Economic Journal* 22, no. 86: 237–47, https://doi.org/10.2307/2221777.
Ruskin, J. (1860), 'Earth-Veil', in C. Tinker (ed.), *Selections from the Works of John Ruskin*, Cambridge: Riverside Press, https://www.gutenberg.org/files/15200/15200-h/15200-h.htm#Earth.
Seizure (2008), intervention, London: Artangel/Jerwood, 151–189 Harper Road, 3 September to 30 November 2008.
Senda, S. (2018), 'Archaeology of the Future: Tsuyoshi Tane opens two exhibitions in Tokyo', *Designboom*, 12 September 2018, https://www.designboom.com/architecture/tsuyoshi-tane-tokyo-exhibitions-archaeology-future-12-09-2018/.
Showing Making: International Conference on the Representation of Image Making and Creative Practices in Ritual, Art, Media and Science (2009), conference, Amsterdam: Dutch Utrecht University, Royal Netherlands Academy of Arts and Sciences, Dutch Filmmuseum, 18 to 19 June 2009.
Silver, M. and S. Burns (2018), *Barron and Larcher: Textile Designers*, Woodbridge: ACC Art Books.
Somers, T. (2015), 'The lengths to which artists go: Matthew Barney at MOCA', in *200-percent*, https://200-percent.com/matthew-barney-at-moca/.
Space Shifters (2018), exhibition, London: Hayward Gallery, 26 September 2018 to 6 January 2019.
The Story of Japanese Textiles (2019), exhibition, Kyoto: Hosoo Gallery, November 2019 to February 2020.
Storey, J. (1974), *A Manual of Dyes and Fabrics*, London: Thames and Hudson.
Storey, J. (1978), *A Manual of Textile Printing*, London: Thames and Hudson.

Sudo Reiko: Making NUNO Textiles (2019), exhibition, Hong Kong: CHAT, the Centre of Heritage, Arts and Textile, 24 November 2019 to 23 February 2020.

Sudo Reiko: Making NUNO Textiles (2019), online video, Hongkong: CHAT, the Centre for Heritage Arts and Textile, https://www.youtube.com/watch?v=L1RF8l5Bu2Q&feature=emb_logo.

Tamashige, S. (2018), 'Tsuyoshi Tane designs forward-thinking buildings by building on memories', *Japan Times*, 4 November 2018, https://www.japantimes.co.jp/life/2018/11/04/general/tsuyoshi-tane-designs-forward-thinking-buildings-building-memories/.

Tanaka, C. (2019) , interview, Kyoto: Gallery Gallery, 5 November 2019.

Tane, T. (2019), *Archaeology of the Future*, lecture, London: Daiwa Anglo-Japanese Foundation, 9 October 2019, https://dajf.org.uk/event/archaeology-of-the-future.

Tansa: Japanese Threads of Influence (2022), exhibition, Farnham: Crafts Study Centre, 4 January to 26 March 2022; Kyoto: Gallery Gallery, 23 April to 8 May 2022.

Tavernier, J. B. (1925), *Travels in India 2*, London: Oxford University Press.

Teshima Project (2019), exhibition project, Japan: Teshima, https://at-ta.fr.

Textile Magicians (1997), online video, dir. Cristobal Zanartu, New York: Mystic Fire Video, https://vimeo.com/139602030.

The Textiles Collection – University for the Creative Arts at Farnham, https://www.vads.ac.uk/digital/collection/ST.

The Theatricality of Chiharu Shiota's Art: Brilliant Ideas Ep. 52 (2017), online video, Bloomberg Quicktake, 1 May 2017, https://www.youtube.com/watch?v=ULpxdHy0eZc.

Toerin, E. S. (2003), 'Mud cloth from Mali: its making and use', *Journal of Family Ecology and Consumer Sciences* 31: 52–7, https://www.ajol.info/index.php/jfecs/article/viewFile/52840/41441.

Touchon, A. (1959), 'Note sur la teinture des laines au Djebel Amour', *Cahiers des Arts et Techniques d'Afrique du Nord* 5: 118.

Traditional Amami Mud Dye (2018), online video, Japan: Kanaikougei, http://www.kanaikougei.com/#koubou.

Tuan, Y. (1975), 'Place: An Experiential Perspective', *Geographical Review* 65, no. 2: 151–65, https://doi.org/10.2307/213970.

Tuan, Y. (1977), *Space and Place: The Perspective of Experience*, Minneapolis: University of Minnesota Press.

UNESCO (2003a), *Text of the Convention for the Safeguarding of the Intangible Cultural Heritage*, https://ich.unesco.org/en/convention.

UNESCO (2003b), *Traditional Craftsmanship*, https://ich.unesco.org/en/traditional-craftsmanship-00057.

UNESCO (2003c), *What is Intangible Cultural Heritage?*, http://www.unesco.org/culture/ich/en/what-is-intangible-heritage-00003.

Unger, M. (1992), 'Chiyoko Tanaka', in *Kohyama and Tanaka: Contemporary Ceramics and Textiles Art from Japan*, 28–9, Leeuwarden: Museum of Princessehof.

The Village, webpage, https://juntomita.com/the-village/.

Visual Arts at Sadler's Wells (1999), *Anya Gallaccio: All the Rest is Silence*, London: Sadler's Wells.

Vydra, J. (1954), *Indigo Blueprint in Slovak Folk Art*, Prague: Artia.

Wada, Y. (2002), *Memory on Cloth: Shibori Now*, New York: Kodansha.

Wada, Y., M. Kellogg Rice and J. Barton (1985), *Shibori: The Inventive Art of Japanese Resist Dyeing*, New York: Kodansha.

Weir, J. (2008), *Walking the Block*, Matlock: Templar Poetry.

Wiles, W. (2008), 'Review: Seizure', *Icon*, https://www.iconeye.com/news-parent-category/review-seizure-2.

Wilson, R. (2003), 'The birth of a notion: Richard Wilson on the genesis of 20:50', *Guardian*, 4 April 2003, https://www.theguardian.com/artanddesign/2003/apr/04/thesaatchigallery.art3.

Yokobori, S. (2016), 'Living in the Mashiko Milieu', in *Inspired by Mairet: Higeta Tadashi and Ethel Mairet*, 30–2, Mashiko: Committee for Promotion of Mashiko Culture Town.

Yoshida, S. (2023), letter, 17 April 2023, trans. R. Ohkanda.

Index

Page numbers in **bold** refer to figures.

actions, 2–3
Adult Education Centre, Dartington, 12
African printing, 17–8, 81–2
Algeria, 22–3
All the rest is silence (Gallaccio), 92, 97–9, **98**
Amami Oshima, 82–3
Amazonian Indians, 29
Amsterdam, 1–2
aniline dye, 8
applied dye, 54
Arimatsu, Japan, 52–3, 167–9
Artangel, 100, 102
Arts and Crafts movement, 24–5
Atelier Tsuyoshi Tane Architects website, 119
Austria
  im.material exhibition, 140–3, **142**, **143**, **144**, **145**
  Steinberg, 131, 133, **134**, 135, 138, **140**
Avon, River, 86–7

Balfour-Paul, Jenny, 18, 20–1, 22–3, 28–9, 30–1, 32–3, 36, 67, 69–70, 72–3, 76
Bancroft, Edward, 7–8, 22, 67, 73
Barnes, Ruth, 16–7
Barney, Matthew, 91, 102–4
Barron, Phyllis, 10–1, 40–1, 43–4, 48, 52
Barton, Jane, 15, 16, 20, 25–6, 26–7, 45, 52–3, 92, 93, 168
BASF, 8
batik, 9, 17, 34, 81
Baume, Nicholas, 78–9
Belsey, Faye, 19
Berlin, 119–20
Berlin Wall, 119–20
Bhabha, Homi K., 3, 3–4, 79, 92, 102–3, 105–6, 111–2, 138, 139, 143, 172–3
binding, 19–20
black and blackness, 23–4, 67, 80–1, 107
Blake, William, 85
bleaching agents, 82
block printing, 12, **13**, 14, 29–30, 48–9
  rhythm, 49–51
blue and blueness, 67, 69
Blue Comes from Indigo, Yet it is Bluer than Indigo Itself (film), 70, 73–5, 146, 169–70
blueprint, contemporary, 138–9
bodily action and engagement, 2–3, 92–7, 111–2
bogolan, 18, 81–2
bokashi, 54
Bosence, Daniel, 29–30
Bosence, Susan, 11–2, 21, 25, 27, 48
  Hand Block Printing and Resist Dyeing, 12, 14, 21–2, 29–30, 40, 51–2
  practice, 41–3, 44–7, 49
  Resist Patterned Length, 44, **44**
  workspace, 40–3
Boundary (Brassington), **160**
bronze, 102
Browngrotta Arts, 97
buildings, 3–4
burnishing, 76–8
Butler, Robert, 86–7

calendering, 135, 138
calico printing, 81
Callender, Jane, 41
Camberwell School of Arts and Crafts, 12
Cardon, Dominique, 18, 23–4, 28, 30–2, 32–3, 67, 69, 70, 72
Centre for Folk Art Production (ÚĽUV), 15
chemical resist prints, 17
China, 18–9, 70–1, 77–8
clamping, 9

The Collection as a Narrative in the Crafts (Brassington), 10
colour, 35–7
  material approaches, 67
  stuff of, 67, 78–81
Colour and the Calico Printer: An Exhibition of Printed and Dyed Textiles 1750–1850, 12
Colour into Cloth exhibition, 21
colouring agents, 76
communication, 4
Compressed (Brassington), **168**, 169
Crafts Council, 21, 25
Crafts Study Centre, 10, 11
Crafts Study Centre, 2002.33 (Brassington), 50–1, **50**
creative entanglements, 170–1
creative opportunity, heritage as, 173
Cresswell, Tim, 41–2
Crosland, Neisha, 48–9
cultural exchange, 17
cultural heritage, 3
cutting tools, 49
Czechoslovakia, 15–6, 34, 35, 117

dabu, 81, **82**
Danglová, Olga, 15–6, 34–5, 105
Dartington, Adult Education Centre, 12
Dartington College, 11
de Waal, Edmund, 30
design, 12, 48–9
Detschmann, Gabriele, 140
Dezeen, 165
Diagonal (Brassington), **159**
Ditchling Museum, 26
domestic workshop, the, 40–3
dorozome dyeing, 83
Dorrell, Dan, 119
Dress (c. 1815–20), **13**
drying, 131, **132**
drying cabinets, 57
drying racks, **60–1**
dye vats, 22, 103, 111, 112, **128**, **129**, 131
dyed paper, **62–3**
dyeing
  as performance, 22
  physicality, 23
  process, 112

The Dyer's Handbook: Memoirs of an 18th Century Master Colourist, 23–4, 28, 31

Edo Komon workshop, Tokyo, 164–5
Egyptian papyrus inscription, 8, 22, 33
Ellipse 2 (Brassington), **161**
embodied skill, 3
emotional experience, 67, 74–5, 104–12
Encrusted (Brassington), 88, **89**
Enlightenment, Age of, 31
Entangled (Brassington), **174**
environmental perception, 3
Exeter County Hall, 46
The Exhibition of the Works by Linda Brassington, 112

Farnham, University for the Creative Arts, 10–1, 11, **13**, 15, 59, 70–1
Farnham School of Art, 11–2, 14–5
Farther Memory (Shiota), 119–20
Faulkner, Rupert, 109–10
finishing, 77–8
fixing, 76
folding and clamping, 19–20, 167
France, 70
Fukumoto, Shigeki, 35
Fukumoto, Shihoko, 54, 91, 93–5, **94**, 102–4, 109–11
Fukumoto Shihoko: Japan Blue exhibition, 109–10

Gallaccio, Anya, 91, 92, 97–9, **98**
Garfield, Simon, 8
Gayford, Martin, 109
Germany, 76
Ghotmeh, Lina, 119
Gittinger, Mattiebelle, 16
Globe Theatre, 76
Granger-Taylor, Hero, 12, 14
Guardian, 32, 80, 87–8, 107
guidelines, 49

Hamada, Shoji, 147, 151
Handmade in Japan: Mud Dyeing the Silk (TV programme), 83
Hara, Kenya, 35–7
Harisho family, 167–9
heritage, 118

as creative opportunity, 4, 173
definition, 139
heritage sites, 3
Higeta, Hiroshi, 150
Higeta, Tadashi, 55, 59, 146, 148, 151–2
Higeta Aizome Kobo, 88, 112, **113**, **114–5**, 147, **148**, 148–53, 169–71, **171**, **172**, **173**, **174**
indigo vats, 54, **62–3**, **68**, 109, **108**, **116**, **149**, **154–6**
shibori process, 54, 55–9, **60–1**, **62–3**, **64**, **65**
stencil dyeing, 54, 55–9, **60–1**, **62–3**, **64**, **65**, 150, 151, **157–8**
Hiorns, Roger, 91, 99–102, **101**, 105
Holburne Museum, Bath, 11
Homme Plissé Issey Miyake, 165–6, **166**, 167, **167**
Hosoo, Masataka, 120
Hosoo Gallery, Kyoto Prefecture, 120, **121**
Hranovnica, Slovakia, 125–7, **126**, **127**, **128**, **129**, **130**, 135, **136–7**
humidity, 164

Icon, 100–1
ikat, 9
im.material exhibition, 140–3, **142**, **143**, **144**, **145**
Independent, 21
India, 9–10, 16–7, 32, 81, **82**
indigo
dries, 70–1
history, 2, 7
material qualities, 2
preparation, 69–73
price, 9
production, 2, **113**
shades, 69
significance, 9–10, 69
as space, 67
substance of, 67
synthesis, 8, 10
indigo pigment, 76, 76–8
indigo vats, 41, 42, 54, **62–3**, **68**, 73–5, 76, 88, 92, 105, 108–9, **108**, **116**, 141, 145, **149**, **154–6**
indigofera tinctoria, 2
Indigofera tinctoria, 69
Ingold, Tim, 1–2, 3, 39, 46, 54, 55, 67, 79–80, 83–4, 88, 91, 92–3, 93–4, 95, 96, 99, 118, 139, 141, 170
innovation, 117
Inside – Outside (Shiota), 119–20
Inspired by Mairet: Higeta Tadashi and Ethel Mairet exhibition, 148–50
intangible cultural heritage, 1, 3, 5, 152
language of, 33–7
redefining, 4, 171–3
International Biennial of Tapestry 1990, 110
international cooperation, 34
iron, 81, 82–3

Jacob, Mary Jane, 54, 93, 104, 110, 111
James Hockey Gallery, West Surrey College of Art, 12
Japan, 16, 82–3, 120–1
Arimatsu, 52–3, 167–9
indigo preparation, 70
industrialization, 53
Kyoto, 145–7
Tokyo, 119, 164–6, **166–7**
understanding of colour, 35–7
see also Higeta Aizome Kobo; *shibori*
Japan Times, 119
Jeffries, S., 32
John Hansard Gallery, Southampton, 98–9
Jones, Jonathan, 87–8

Kanai, Yukihito, 83
Kanezo, Suzuki, 53
Kapoor, Anish, 67, 78–81, **80**, 92, 106, 111
kasuri, 9
katazome paste, 54, 55–9, **60–1**, **62–3**, **64**, **65**, 152, 153, **157–8**, **159**, **160**, **161**, **163**, 170, **172**
Katsuhiko, Ushiro, 110–1
Kellogg Rice, Mary, 15, 16, 20, 52–3, 92, 93, 168
Kennedy, Randy, 104
King, Brenda, 24
Kitamura, Jim, 70, 73–4
knotting, 19–20
knowledge exchange, 141
Kofun Stadium, 119
Kojima, Takumi, 170
Komeya, Yasumasa, 164–5

Koó, Josef, 131, 133, **134**, 135, **136–7**, 138, **140**, 141, **144**
Kyoto, 145–7
Kyoto Prefecture, Hosoo Gallery, 120, **121**

labour, division of, 52–3
language, 7–12, 18–37
and creation of place, 4
of intangible cultural heritage, 33–7
narrative, 30–3
poetic, 26–7
of space and place, 28–30
visual and sensory, 18–26
Larcher, Dorothy, 10, 11, 25, 27, 40, 43–4, 45, 48, 52
Leach, Bernard, 147
leather, 76
linen, 138
Lingwood, James, 102
literature, 1, 7
intangible cultural heritage, 33–7
international perspective, 15–8
narrative language, 30–3
pedagogy, 11–2, 14–5
poetic language, 26–7
and space and place, 28–30
visual and sensory language, 18–26
Long, Richard, 67, 86–8, **86**
lye, 53–4, 74

Machell, Thomas, 32
Mack, John, 17–8
McLeod, Ella, 11
McQuaid, Matilda, 53–4
Mahtani, Nikhita, 84
Mairet, Ethel, 26, 147
Mali, 18, 81–2
mame shibori, 168–9
mangles, 135
Marx, Enid, 43–4, 48, 49
Mashiko, 92. *see also Higeta Aizome Kobo*
Mashiko Museum of Ceramic Art, 59, **65**, 112, 117, 150, **153**, **162**, 169
masking tape, **64**
mass production, 84
Master Dyers of the World exhibition, 16
material, performativity of, 2
material culture, 79–80
material flow, 97–9, **98**
material transformation, performativity as, 99–102, **101**, 102–4
material transmutations, 91–2
materiality, 1–2, 5, 118
Matsuzaki, Yuko, 150
memory, 33, 36, 93, 117, 118–20, **121**, 126
Millar, Lesley, 96–7, 103, 145
miura technique, 52
Modrotlač: Indigo Country Cloths and Artefacts exhibition, 15
Montško, Elemír, 125–7, **126**, **127**, **128**, **129**, **130**, 135
Moreton, G., 99
Morioka, 73–5
Morris, Michael, 102
Morris, William, 24–5, 92, 104–5
movement
power of, 49
and space, 41–3, 44
mud, 2, 81–3, **82**, 86–8, **86**, 165
mud cloth, 18
Muriuki, James, 141, 142
Myer, George Hewitt, 16

Nai, Manish, 67, 84–5
Nakano, Eisha, 55
national identity, 139
Naylor, Gillian, 24–5, 104–5
netsuke, 30
New York Times, 104
Niedermayr, Lisa, 140, 141, 142
Nigeria, 17, 76

O'Connor, Deryn, 11–2, 14–5, 18–9, 21–2, 28, 40, 41, 44–5, 47, 67, 71–2, 77–8, 81, 85
Ohshima, Seiji, 150–1
Olding, Simon, 40
overdyeing, 131
overprinting, 49
Oxford, Pitt Rivers Museum, 18–9
oxidation, 2, 35, 58, 76, 130, 138, **140**, **154–6**

painted paste resist, 44–7
paper, 76
pattern repeats, 48–9
patternmaking, 12
pedagogy, 11–2, 14–5

peeling, 59
performance, dyeing as, 22
performativity, 5, 91
  as bodily action, 92–7
  of material, 2, 3
  as material transformation, 99–102, **101**, 102–4
  as physical and emotional experience, 104–12
Perkin, William, 8, 10
physical experience, 104–12
Picton, John, 17–8
Pierced (Brassington), 152, **163**
pigment
  indigo as, 76, 76–8
  as stuff, 78–81, **80**
Pitt Rivers Museum, Oxford, 18–9
place, 28–30, 39
  attachment to, 126
  creation of, 4, 152
  definition, 118
  memories of, 118–20, **121**
  power of, 117–8
pole-wrapping process, 53
polishing, 76–8, 85, 135
Powers, Alan, 44, 49
practice, sites of, 3
printing blocks, 48
professional studio, the, 43–4
Púchov, Slovakia, 92, 105, 108, 130–1, **132**, 135, 138
pullies, **128**

record-keeping, 15
registration points, 49, 50, 164
repetition, skill of, 50
reserve patterning, 9
resist dyeing, conceptualisation, 2–3
resist paste, 56, 59. *see also katazome* paste
resist processes, 9
rhythm, 44, 47, 49
  block printing, 49–51
  Higeta Aizome Kobo, 54, 55–9, **60–1**, **62–3**, **64**, **65**
  shibori, 51–4
Rhythm (Brassington), 39, 59, **64**, **65**
River of Fundament (film), 104
Robinson, Stuart, 11, 33
  A History of Dyed Textiles, 8–9, 19–20
  A History of Printed Textiles, 8–9
Roger Hiorns, Seizure 2008/2013 at Yorkshire Sculpture Park (video), 100, 101–2
Roscoe, Barley, 21
routine, 43
Roxburgh, William, 22
Royal Academy, 79
Royal Economic Society, 9–10
ruching, 19–20
Runge, Friedlieb Ferdinand, 10
Ruskin, John, 67, 83, 84

sample strips, 57
Sasaki, Ryuta, 67, 73–5
sasawashi paper cloth, 153, **159**, **160**, **161**, **163**
Second World War, 44
Seiryusui Spring Association, 74
Seizure (Hiorns), 91, 99–102, **101**, 105
Senda, Shuhei, 119
sensory experience, 118
sensory spaces, 4
sharinbai tree, 83
shibori, 16, 20, 33, 34, **51**, 92–7, **94**, 167–9
  miura technique, 52
  pole-wrapping process, 53
  process, 52–4, 54, 55–9, **60–1**, **62–3**, **64**, **65**
  rhythm, 51–4
shimi (deep penetration), 35
shimijimi (keenly felt emotion), 35
Shindigo 23-01 (Shindo), 145–6, **146**
Shindo, Hiroyuki, 53–4, 91, 97, 99, 145–6, **146**
Shiota, Chiharu, 117, 118–9
Showing Making conference, 1
Six Squares, Indigo Blue (Tanaka), 96–7
Slovakia
  Hranovnica, 125–7, **126**, **127**, **128**, **129**, **130**, 135, **136–7**
  Púchov, 92, 105, 108, 130–1, **132**, 135, 138
Soledons, 41
Somers, Thierry, 102, 103
space, 28–30, 39, 108, 109–11
  domestic, 40–3
  indigo as, 67
  and movement, 41–3, 44

professional, 43–4
sense of, 127, 130
significance of, 55
Space Shifters exhibition, 106–8, **106**
Spectator, 109
staining, 76
Steinberg, Austria, 131, 133, **134**, 135, 138, 139
stencil dying, 54, 55–9, **60–1**, **62–3**, **64**, **65**, 150, 151, **157–8**, 164–5
Stonehenge, 86
Storey, Joyce, 17
The Story of Japanese Textiles exhibition, 120
storytelling, 30–3, 117
Sudo, Reiko, 35, 117, 122–5
sukumo, 70, 73–4
Surrey Nano Systems, 80

tactile sensibility, 139
Tanaka, Chiyoko, 67, 85–6, **87**, 88, 91, 95–7
Tane, Tsuyoshi, 117, 119–20, 146
Tanner, Robin, 11
tannins, 2, 81, 82, 83, 164
Tansa: Japanese Threads of Influence exhibition, 145–6, **146**, 169
Tavernier, J. B., 22–3
Tea Room – Mist (Katsuhiko), 109–11
tensioning, **136–7**, 138
tenugui, 168
Textile Magicians (film), 97
Textile Museum, Washington, 16
textile thinking, 7
Textural Space exhibition, 94–5
theatres of making, 1
tie and dye, 9, 54
time, 44, 86, 152
Time magazine, 52
tjanting, 50–1
Tokyo, 119, 164–6, **166–7**
Tomita, Jun, 146–7
Touchon, A., 22
Trnka, Peter, 131
Trnka, Stanislav, 49–51, 105, 130–1, **132**, 135, 138
Trnková, Helena, **132**
Tuan, Yi-Fu, 3, 4, 39, 41, 41–2, 42–3, 44, 46, 50, 53, 117, 118, 126, 133, 152
Turner Prize, 102
20:50 (Wilson), 92, 106–8, **106**

Uchiyama, Takeo, 86, 88
Unger, Marjan, 95, 95–6, 96
United Nations Educational, Scientific and Cultural Organization (UNESCO), 3, 5, 15–6, 33–4, 143, 152, 173
University for the Creative Arts, Farnham, 10–1, 11, **13**, 15, 59, 70–1

Vantablack, 80–1
vat dyes, 41
Vienna, *im.material* exhibition, 140–3, **142**, **143**, **144**, **145**
Void (Kapoor), 79
Vydra, Josef, 15, 34, 35

Wada, Yoshiko, 15, 16, 20, 33, 52–3, 92, 93, 168
Wagner, Karl, 141, 142
Wardle, Thomas, 24–5
Washington, Textile Museum, 16
water, 28, 74
Water Castings (Barney), 91, 102–4
wax, 46–7
wax resist prints, 17, 19, 46–7
weaving, 151
Weir, Jane, 26–7, 28
West African Prints, 17
West Indies, 10
West Surrey College of Art and Design, 11, 12, 15
Wiles, William, 100–1
Wilson, Richard, 92, 106–8, **106**
woad, 70
woad paint, 76
workspaces, 4, 103
domestic, 40–3
last surviving in Europe, 138–9
professional, 43–4
writing styles, 4

Yemen, 23, 28–9
Yokobori, Satoshi, 151
Yoruba people, 76
Yoshida, Shigeru, 73–5, 74–5
Yoshioka, Tokujin, 165